LIFEWORLD AND TECHNOLOGY

Current Continental Research is co-published by The Center for Advanced Research in Phenomenology and University Press of America, Inc.

CURRENT CONTINENTAL RESEARCH 009

LIFEWORLD AND TECHNOLOGY

Edited by

Timothy Casey
and
Lester Embree

1990

Center for Advanced Research in Phenomenology
& University Press of America, Washington, D.C.

University Press of America,® Inc.

4720 Boston Way
Lanham, MD 20706

3 Henrietta Street
London WC2E 8LU England

Printed in the United States of America

British Cataloging in Publication Information Available

Co-published by arrangement with
The Center for Advanced Research in Phenomenology

Library of Congress Cataloging-in-Publication Data

Lifeworld and technology / edited by Timothy Casey and Lester Embree.
p. cm. — (Current continental research ; 009)
Includes bibliographical references.
1. Technology—Philosophy.
I. Casey, Timothy. II. Embree, Lester E. III. Series.
T14.L49 1989 601—dc20 89–39437 CIP

ISBN 0–8191–7626–5 (alk. paper)

The paper used in this publication meets the minimum requirements of American National Standard for Information Sciences—Permanence of Paper for Printed Library Materials, ANSI Z39.48–1984.

TABLE OF CONTENTS

Introduction vii

Living In Electronic Space
Langdon Winner 1

Nazi Biomedical Technologies
Robert N. Proctor 17

Technology and Liberation
Algis Mickunas 49

From Recorded Speech to the Record of Thought
Ivan Illich 69

Text and Things: Holding on to Reality
Albert Borgmann 93

Literacy, Mediacy and Technological Determinism
Larry Hickman 117

Media as Lifeworld
Wolfgang Schirmacher 133

A Perspective on the Rationality of Scientific Technology or How to Buy a Car
Lester Embree 145

Technology and the Objectivity of Values
Joseph C. Pitt 165

Patterns of Discovery or Social Construction of Technology: The Invention of the Turbojet
Edward W. Constant II 181

Architectura Architecturans: World(s) in the Making
Ronald Bruzina 201

On Going to Church and Technology
Carl Mitcham 221

The Religious Dialectic and Technology
Frederick Ferré 237

Designing Excellence: Some Functional and Aesthetic Considerations
Timothy Casey 243

Translating Technology: From Artifact To Habitat
Gayle L. Ormiston 259

Sartre and Technological Being-In-the-World
Thomas R. Flynn 271

The Hidden Life of Technological Artifacts
John J. McDermott 289

Notes on Contributors 303

Index 307

INTRODUCTION

A perplexing question for philosophers today is why technology has received so little philosophical attention. There is simply no denying the facts that technology in one form or another has always been with us, that we now live in a radically changing technological culture and age, and that our images and conceptions of ourselves are increasingly reflecting the technological character of our environment. To say that technology is what makes us human may be going too far; to say that we can be human without technology is absurd. Why then have philosophers apparently neglected such an important dimension of human existence?

Before addressing this question, however, one must admit that it is slightly skewed. For there exists, in fact, a rich, albeit hidden, history of reflection on matters technological that stretches back through Western philosophy to Aristotle and Plato.[1] The concept of *techne/arts*/technique has always played a role in ancient, medieval and modern thought. Subtract the concern with technology from, say, Marxism and Pragmatism and they disappear.

Nevertheless, what has come to be known as the "philosophy of technology" is a small movement of relatively recent origin. Although it encompasses virtually all contemporary schools of philosophy, it does not run very deep in any one of them. (Significantly, this has led philosophers of technology to put aside their parochial philosophical differences for the sake of making common cause.) It is therefore puzzling that an issue that pervades our lives and will undoubtedly come to dominate our headlines and politics more and more has stimulated so little philosophical interest and discussion.

One answer to this oddity of contemporary intellectual life has to do with a certain common-sensical and rather pervasive way of perceiving and thinking about technology. Technology, it is alleged, is merely a neutral tool, a human invention that can be used for good or ill depending upon the intentions and competences of its human users. This is a more sophisticated version of the current saw "Guns don't kill people, people do." Hence, it is concluded, there is nothing intrinsically interesting about technology from an ethical, political or any other non-technical point of view. Not too consistently, we tend to think technology is an "obvious" good that varies in its consequences depending upon the society in which it is embedded and the social uses to which it is put. To question its beneficence is to question civilization itself.

Now the neutrality thesis concerning technology–while it takes us in the direction of an answer to our question–itself needs further clarification and explanation. The root of the instrumental conception of technology is to be found in the modern link between science and technology, a link usually expressed by defining the latter as "applied science" in spite of massive historical evidence to the contrary.[2] Such a misconception can only be explained by pointing to the dominant status modern science has achieved in our culture. Why *this* is so cannot, of course, be taken up here. Instead, we offer a brief phenomenological account of how science has come to obscure the significance of technology and how phenomenology has served to counteract that obfuscation.

As Husserl made clear in the *Crisis,*[3] the world in which we live is not the objective world of nature presupposed in mathematical physics. Science does not give us the lifeworld, but only a construct that can be known but not experienced.[4] Thus, while matter-in-motion may be "true," it is hardly fundamental so far as what is disclosed by classical phenomenological method and description. This is not to say that nature is not to be found in the lifeworld. It is. But it is not the "nature" studied by the natural sciences. Interwoven with experienced nature is a social and cultural reality that complicates (and enriches) the picture considerably.

As an intentional correlate of human consciousness, the lifeworld presents itself in terms of a variety of modes of givenness of objects and horizons. Broadly speaking, this variety consists in moral, political, aesthetic, religious, linguistic and historical dimensions–all of which make up what Husserl calls the lifeworld's "concrete and infinite historicity." Furthermore, it was the virtue of Heidegger, Husserl's student, and others who followed in his wake to correct an imbalance in Husserl's account by emphasizing the *practical* character of the lifeworld.[5] It was in this way that the scientific hold on modern philosophy was broken and technology brought into the light as worthy of study and reflection in its own right, apart from science and its methodologies.

For phenomenology, then, there is a distinction between the natural universe and the lifeworld. The lifeworld is fundamentally a practical reality or situation in which we live and have our various projects and concerns. Whatever sense and coherence this world might have is initially constituted at a pre-reflective level of existence that is primarily "technological." It is thus through the straightforward exercise of bodily skills in tool and equipment use that the horizon of the world is grasped and made explicit. And it is here that the practical character of this world is seen as opening out onto the social, aesthetic, political and ethical dimensions that comprise its richness and complexity.

The concept of the lifeworld, however, involves more than the delimiting of the epistemological claims of modern science and the *theoretical* opening up of a realm correlative to the full range of human experience. It also serves as a way within the modern context to connect the moral and political concerns of everyday life with technologies that until now were either ignored philosophically or believed to be neutral with respect to traditional, *practical* questions of philosophy concerning wisdom and goodness. To situate technology in this way is no easy task. Nor can one say that it is a completed one.

The motivation behind this volume springs from a sense of this task, even though the orientation of many of the contributors is not phenomenological. Originally most of these essays were presented at the conference entitled *Lifeworld and Technology* held by the Center for Advanced Research in Phenomenology, Inc. at Duquesne University in the Fall of 1987. Its purpose was two-fold: to demonstrate the embeddedness of technology in everyday life (especially the 20th century lifeworld), and to pursue the broader philosophical ramifications and questions raised by this embeddedness.

In bringing together philosophers of various stripes (pragmatist, phenomenological, existentialist, deconstructionist, analytic, hermeneutic) with human scientists of varying interests (history, political science, linguistics), the conference focused on the issue of technology and not on the academic and philosophical differences of the participants. As a result, the range of topics—which extends from turbojets and biomedical technology through architecture to reading and politics—matches the breadth of philosophical and scientific approaches represented. Thus one can begin to see in a book like this that technology does not belong to the province of just one philosophical school, or even to philosophy itself. It is complex and deep enough to transcend the bounds of any one method of research and of any particular set of philosophical presuppositions.

Several themes emerged from the three days of papers and discussion and are reflected in this collection. First, the conference and this book are proof that a pluralistic approach can work in the field of technology studies. Although agreement was rare, stimulation, insight and a common sighting of problems were the order of the day. In short, tying the participants together was a tacit conviction that technology *is* a philosophical and human scientific issue in its own right and needs to be studied as such. To consider it merely as a sub-branch of the philosophy of science is to miss its historical importance and to distort its genuine philosophical significance.

Second, philosophy of technology can no more ignore 20th century philosophy of science than it can the cultural and historical contexts within which technology develops and has its being. Philosophers of technology have much to gain from interaction with historians, political scientists, sociologists, etc., especially with respect to their understanding of the "worldly" character of technology. The human sciences serve to anchor philosophy in the factual, historical world, and to keep the philosopher close to the "things themselves."

Third, the fruitfulness of an multidisciplinary approach to the study of technology makes it clear that technology extends its reach into every corner of the lifeworld. There is no area of human existence–or the world, for that matter–that is immune from the effects and influences of modern technology. Although there is honest disagreement about the beneficence of technology's intrusiveness into our lives, no longer does one find a scepticism concerning its dominance in our culture. The question, rather, is how and to what extent technology shapes contemporary culture and is shaped by it, and to what extent we can influence this dialectical interplay.

Finally, in considering such questions from different philosophical and scientific perspectives, the participants manifested a marked ambivalence in their descriptions and evaluations of contemporary technological culture, even when emphasizing the necessity and positive value of technology. The deep issue, it seems, is human freedom. For although all are agreed that technology is needed to overcome the exigencies of physical existence, it is not clear to what extent modern technology has come to constitute a new necessity–a "second nature," as it were–and is hence a threat to liberty and those customs, habits and institutions that currently guarantee our newly won freedoms. There appears, in other words, to be an ambiguity establishing itself at the center of our technological culture and calling forth the kind of questioning and reflection typified by the pieces which follow here. Thus no resolution to these problems was forthcoming, nor was a consensus concerning what is to be done about them achieved. It is in this sense that this book is faithful to our current situation and mirrors the complex phenomenon called 'technology.'

We wish to thank all of those who participated in the conference as well as those who contributed new compositions especially for this volume. Also, we extend a special note of gratitude to Eleanore Wentland for her patience and diligence in the typing and preparation of this manuscript and for helping preserve the human element in this difficult technological project. Finally, we thank Duquesne University, Polytechnic University, and the Pennsylvania Humanities Council for financial support for the conference organized at Duquesne University by the Center for Advanced Research in Phenomenology, Inc.

Timothy Casey
Lester Embree

ENDNOTES

[1]See Carl Mitcham, and Timothy Casey, "Notes on the History of the Philosophy of Technology with Some Relations to Imaginative Literature," in Mark Greenberg, and Lance Schachterle, eds., *Literature and Technology* (forthcoming 1990, Lehigh University Press).

[2]The work of Lewis Mumford and Lynn White, Jr. has not only established the historical priority of technology over science, but also has supplied data for asserting an ontological priority as well. See Don Ihde, "The Historical-Ontological Priority of Technology Over Science," in his *Existential Technics* (Albany: State University of New York Press, 1983), pp. 25-46.

[3]Edmund Husserl, *The Crisis of the European Sciences and Transcendental Phenomenology* (Evanston: Northwestern University Press, 1970).

[4]Phenomenology goes further and asserts that science is constructed upon the lifeworld and the kind of experiences that serve to ground the formalism of any mathematically oriented knowledge. Merleau-Ponty says it best in the *Phenomenology of Perception*: "All my knowledge of the world, even my scientific knowledge, is gained from my own particular point of view, or from some experience of the world without which the symbols of science would be meaningless. The whole universe of science is built upon the world as directly experienced" (p. viii).

[5]That Husserl recognized a "natural primordial attitude" that is practical and prior to science and philosophy is evident in the Vienna Lecture of 1935, "Philosophy and the Crisis of European Humanity," in *Crisis*, loc. cit.

LIVING IN ELECTRONIC SPACE

Langdon Winner

Abstract: Contemporary developments in microelectronics and telecommunications create a tension between electronic information and spatial location. A person or institution may exist not only as a physical presence in a particular place, but also as a disembodied entity within electronic space. Conventional moral and political theories assume that communities and citizen roles are spatially defined, rooted in specific pieces of land. If this assumption is no longer valid, how should theorists respond?

Where am I?

The question suggests disorientation. The writer must be lost. Someone ought to help him regain his bearings.

Where are we? Again the question seems merely confused. To raise it in a philosophical context seems odd. Answers to the question "where" are usually understood to be fairly simple. Look around you. If necessary, ask your neighbor. If there is still some puzzlement, consult a reliable highway map. Surely, it cannot be all that difficult to establish spatial location of something. The truly challenging issues in Western thought have to do with "why?", "how?", and "on what rational basis?", not issues about "where?"

But there is good reason to suppose that in the late twentieth century some important questions about spatial location and its significance no longer have obvious answers. Particularly if one asks where we are when we engage in electronically mediated social interactions, the topic takes on

---

Lifeworld and Technology, eds. Timothy Casey and Lester Embree, Washington, D.C.: The Center for Advanced Research in Phenomenology and University Press of America, 1989.

new interest. My purpose here is to begin a reconsideration of this matter, arguing that our sense of "where" we are will likely have much to do with "who" we may eventually become.

Let us begin with two stories that may help illustrate the new situation. It is March 1987. Pop musicians Stevie Wonder, Nile Rodgers and their colleagues gather together for a recording session. As the music begins, Stevie Wonder plays his harmonica in that distinctive style known to listeners since the early 1960s. Rodgers adds a gentle refrain on guitar while master producer Quincy Jones directs the background musicians and sound technicians. All seems to be going well. An interesting feature of the date, however, is that the two key musicians are not in the same room or, for that matter, even in the same city. Wonder stands at a microphone in a recording studio in Los Angeles with Jones at his side. Nile Rodgers is playing his instrument in a Queens studio, 3,000 miles away. All of them are connected by a system spanning 45,000 miles that includes mixing boards, the Sony 1630 digital sound processors, mobile satellite dishes, the G-Star satellite, and fiber optic cables. The signal arrives in Los Angeles 520 milliseconds after the master recording is made in New York and vice versa, while the original on each end is itself electronically delayed by the same 520 milliseconds; thus, the musicians hear the two sound sources simultaneously. Technical arrangements of this kind create what is in effect a single "recording studio" from geographically separate entities.[1]

But where is the recording session? In Los Angeles? New York? The stratosphere?

Here is a somewhat similar case. A professor at a university in the Northeast leaves his regular Monday evening seminar and drives home. He goes into his study, sits down at his computer and logs into yet another seminar. This one includes participants from various corners of the U.S., South America, Europe, and Asia. The teleconference does not occur in "real time" but rather in a synchronous arrangement that allows participants to tune in at different times to join a conversation that has continued for several months.[2]

But, once again, where is the seminar? And where are those who take part in it?

Situations like these are no longer extraordinary; they have already become commonplace for a great many persons who live and work in the technologically centered environments of the late twentieth century. In such settings the development of micro-electronics and photonics creates a tension between electronic information and spatial location. To a certain extent the new technologies eliminate the importance of geographically defined places altogether. The microcosm of electronic devices sustains a macrocosm of worldwide communications and organizational activities. For those individuals and corporations able to take advantage of this situation, the important horizons become global ones. Production is now an increasingly transnational matter in which firms manufacture and assemble what they need in "instant factories" able to provide the lowest overall cost. Markets for goods, services, and securities are now global entities as well. To take advantage of these new relationships, the management of large firms is predicated on the successful maintenance of multinational, multicultural operations and on the development of personnel able to handle diverse business contexts. New ways of organizing corporate research call for transnational consortia that assemble scientists and engineers from many firms and nationalities together in a manner expected to increase the productivity of R&D. Whatever the organizational activity may be, it is ideally to be placed where there is the greatest comparative advantage, even if that advantage is a fleeting one. Locating a factory, office, laboratory, shopping center, or even company headquarters in a particular place becomes an expediency of the moment, much like having an outlet in which to plug an electrical appliance. In contrast, the industrial age with its cumbersome communications and heavy equipment tended to create productive enterprises in more or less permanent locations—the textile mill, the steel plant, the automobile factory: Manchester, Pittsburgh, Detroit and the like. Nothing in the material conditions of the information age requires any such anchor.

Conditions of this kind are not entirely new. A willingness to transcend spatial boundaries has been a feature of capitalism from its earliest manifestations; profits are sought wherever they are best achieved, regardless of place. From that vantage point, the new technologies merely enhance the centuries-old motivations of bankers and commercial interests. In addition, the spatial extension of personal communication and organizational structures has long been a fact of life, even before the advent of the telegraph and telephone. In the mode of oral communication, one's utterances could be remembered and repeated in a great many places throughout time, as were the Homeric poems in ancient

Greece. With the development of handwriting and, much later, printed communication, a person's words and ideas could be represented in a limitless variety of spatial and temporal settings. Of course, with the coming of electrical and electronic technologies of the nineteenth and twentieth centuries, the range of possibilities for spatially extended communication have increased rapidly along many dimensions.

In the customary view, such abilities are given the general name "communication"– the sending and receiving of messages, from one place to another. A person originates a message at one point and time. The message is then transmitted through some means–paper, record, tape, radio, television, etc.–to be received at some different location. Indeed, much of what we do with such media can still be adequately described in this conventional way; people use the new technologies to communicate with each other.

A significant turning point comes, however, when persons, organizations and events begin to exist as complex, fully featured entities within self-contained electronic boundaries. In this context, the presence of a person or institution may be understood not only as a physical presence situated within a spatial location, but also as a presence or collection of presences situated within electronic space. Actions can be seen not only as deeds that occur at point A and are then transmitted to point B, but more accurately as actions initiated within electronic contexts themselves. In the same manner, whole organizations (or significant parts of organizations) begin to exist as social structures embodied within computer software and other protocols of electronic signal processing. Here the notion of "communications" no longer expresses the conditions that matter. One must find ways to talk about the beings, activities, relationships, and structures that exist within these new spaces. To an increasing extent, the settings in which people live and work are electronic environments that require no bodily presence or physical movement. What are in effect whole rooms, buildings, streets, highways, and cities can be formed of computers, fibre optic cables and telecommunications ports. Conversations and meetings take place within the structures of teleconferencing software. Complex decisions are made through the intersection of streams of telecommunications data. World-shaking events unfold through the confluence of digital signals interpreted by computer programs.

A useful metaphor for such global electronic spaces is offered by the representational signs used in office automation. The goal of such systems is to do away with such material encumbrances as rooms, desks, file cabinets, file folders, paper, and, ultimately, people and to make the information formerly contained in such settings available instantaneously to the eye. Operations on a computer display replace the physical space and social movement of the office. In some computerized representational schemes of this kind, physical apparatus is replaced by little pictures of file cabinets, folders and documents commonly called "icons." When a person wants a particular piece of information, he/she directs an arrow to the relevant icon, pushes a button and the task is done. No longer is there any need to move physically through a world of people and material things; one does not, for example, get up from a desk, walk to a file cabinet, get a piece of paper, go back to one's desk, and sit down. This sequence of motions is replaced by a few simple manipulations on a keyboard, mouse or other device handling the machine/human interface. Computerized information systems of this kind remove action from the body, delivering it to the eye and intellect.

An iconography of a similar sort can now be employed to map the relationships, organizations and events of the era of high-speed digital communications. Within computer based networks linked by satellites, separate nodes interact easily over vast distances. Those who have access to the network's hardware and software enter an electronic lifeworld filled with offices, meeting rooms, mailboxes, bulletin boards, gateways, corridors, command centers, storage areas, repair shops, banks, police stations, even coffee houses, all composed of electronically produced images instantaneously available to the eye. Taken as a whole, such institutions present a physically disembodied social order that exists parallel to and eventually competes with the society of bodies in spatial location. To be enfranchised in this order means establishing one's competence and finding a legitimate role within one or more of the established networks.

An example of the power that such institutions already command can be found in the network of electronic currency markets and stock markets that now engage in trading virtually non-stop, twenty-hours a day. Within this intense, fast-paced, ethereal environment, banks and investment firms speculate, for example, on fluctuations in the value of the yen, dollar, mark, pound, or franc, making or losing millions on a slight turnaround of a currency's price. Traders who participate in such markets stare at computer terminals that supply information on second-to-second changes in price and then signal their sales and purchases through brief messages over telephones. In some cases, the "agents" involved in trading are actually computer programs based on intricate formulas that give orders to buy or sell. Through processes like these, the "wealth of nations" ebbs and flows in ways largely divorced from the traditional, spatially centered arrangements that formerly produced and controlled economic value. The economy of abstract symbols becomes decoupled from the economy of material processes. As represented in the world of electronic information, things ultimately achieve their worth as blips on a display terminal. Everything else becomes an illusion. To believe that actors who move and shake within this context necessarily respect the value of any spatially located accomplishment is, to an increasing extent, a serious misconception. Dwellers on the land in the cities eventually confront the unhappy and perplexing discovery that the old farms, factories, offices, and shops they know to be crucial to their sustenance are suddenly marginal because actors in the network judge them deficient.

This fate may await even those nations that appear to be winners in the race to fashion the most advanced forms of production, communication and exchange. Japanese social planners, for example, believe they can contain the social vitality, economic benefits and political control of high technology development within the boundaries of the Japanese archipelago. The nation's Ministry of International Trade and Industry (MITI) has devised a number of programs of scientific, technical, corporate, and community development to achieve this goal. One aspect of the plan, the "Technopolis" strategy, involves creating more than a dozen new urban centers devoted to innovations in high technology. Certainly, the strong cultural solidarity of the Japanese people enhances their prospects for success. But insofar as the proposed transformation aims at achieving global extension and relies upon advanced computing and telecommunications, it is unreasonable, or at least old-fashioned, to expect that the institutions, products and control of any future Technopolis will remain rooted in any particular piece of island lawn.[3]

The new circumstances presented by electronic globalism and the disembodied agents who inhabit electronic space suggest a need to reconsider our views of political society and all that political societies have heretofore sought to achieve. Most of modern political theory assumes a geographical context for the realization of the fundamental ends it addresses. Our theories proceed by justifying universal norms which could, in principle, be applied anywhere. In practice, however, such norms are applied with the laws and institutions of particular spatially defined communities: modern nation states. Until recently this has seemed a perfectly reliable context for thinking about politics. One asks how shall we, members of a community living on a given piece of turf, arrange our basic institutions?

This is certainly true for the tradition of social contract thinking that runs from Hobbes to Locke, Rousseau and John Rawls. Locke's conception of government, for example, begins with the idea that each man living in a natural condition has "a *Property* in his own *Person*." A man extends that property by mixing his labor with nature, taking things like apples, acorns and sections of land out of the Common. "The *labor* that was mine, removing them out of that common state they were in, hath *fixed* my *Property* in them." In Locke's view, government derives from the need to protect the property of individuals, with property seen first and foremost as a piece of land and its natural bounty. The *Second Treatise of Government* advises that if a person does not care for the arrangements agreed upon at the establishment of government, he is entitled to move elsewhere, that is to another geographically situated place beyond the boundaries of his former society. As Locke puts it, "he is at liberty to go and incorporate himself into any other Commonwealth, or to agree with others to begin a new one, *in vacuis loci*, in any part of the World, they can find free and unpossessed."[4]

Visionaries have occasionally gone beyond such definitions of sovereignty to propose a world government granting citizenship to all inhabitants of the planet regardless of location. But the two modern attempts to create the foundations of worldwide political order–the League of Nations and United Nations–have never, as their very names indicate, transcended the nation state as the ultimate place of membership. Our basic conceptions of politics are still informed by ideas about physically embodied persons who live in particular geographical locations.

But if it is true that many of the most important institutions now exist electronically and globally, if such institutions begin to command an increasing power–power over production and wealth, power over major social decisions, power over the forces that shape everyday life–what is to become of people's sense of social identity and political association? As such conditions evolve, they move issues to the forefront of the agenda for political thought that have long been far in the background. These include basic questions about the boundaries of community, the terms of citizenship and expectations of loyalty. To talk meaningfully about rights, justice, authority, equality, or even freedom, presupposes that there is a context, a spatial context, for the discussion. But if in fact the spatial situation is itself in question, then there is an urgent need to address the matter. Where are we when we are members of a political community?

The answer used to be something as simple and reassuring as: Here we are, right here. But in some important instances, that will no longer suffice. The power to define communities and recognize membership has to a considerable extent shifted elsewhere, beyond the nation state. The realities of transnational technological and economic processes no longer correspond to traditional notions about political and social being.

We find ourselves in a position similar to that of the citizens and philosophers who lived through the collapse of the ancient city state. From the third century B.C. well into the 4th century A.D. the ability of local communities to be self-governing was destroyed by expanding empires of Alexander, Caesar, Augustus, and those who succeeded them. Formerly, the polis had offered people a context of membership, a sense of who they were as part of a larger whole and an understanding of what was rightfully theirs as a consequence. It provided a framework of laws, offices, political practices, and authoritative decisions. Those fortunate enough to be called citizens were able, if they wished, to experience freedom through participation in government. Thus, Aristotle argued, man is a political animal, that is, a being naturally suited to live in a polis.

But with the coming of the age of empire, that sense of membership, security and self-realization through politics ended. Philosophers of the period, the Stoics, Epicureans, and Cynics, wondered openly about matters that had previously been taken for granted. If the polis is no longer the center of political authority, then how shall we define the whole of which we are members? Answers tended to stress the centrality of enormous containers: nature's infinity, the cosmos, universal reason, humanity as a whole, or the body of Christ. "You are a citizen of the universe," Epictetus advised his contemporaries. Thinkers of this period stressed the need for atomized, isolated individuals to work out the terms of their own salvation on a background of cosmic processes. Political membership itself, however, was often described as something entirely burdensome or futile. Writing about his fellow Christians in the third century A.D., Tertullian observes "For us nothing is more foreign than the [Roman] commonwealth. We recognize but one universal commonwealth, viz. the world . . . I owe no obligation to forum, campus or senate. I stay awake for no public function, I make no effort to monopolize the platform, I pay no heed to any administrative duty, I shun the voters' booth, the juryman's bench . . . ."[5]

In our own time there is as yet no prominent equivalent for the political disillusionment that accompanied the rise of ancient empires. Among academic philosophers there is little effort to confront the economic and technological developments that tend to create global relationships and undermine national sovereignty. Most thinkers still assume that national contexts will be adequate to define the key issues in social and political life. An exception can be found in certain varieties of environmental and ecological thought where some have advanced an almost neostoic view, arguing that we must recognize that we are citizens of the "Gaia", the biosphere as a whole, with rights equal but not superior to those of other species. "Think globally, act locally," a common slogan to express this viewpoint, suggests a strategy that seeks to undermine the authority of the nation in environmental affairs, favoring instead an aggregate of decentralized, worldwide initiatives. While such notions about "deep ecology" do not explicitly address the globalism of digital electronics, they mirror a sense that the basic terms of membership in worldly communities are open to question.[6]

Where one can find the beginnings of an attempt to examine the transcendence of spatial identity is in a context where such conditions already pose genuine practical problems: deliberations about international business and corporate management. Thus, political economist Robert Reich wonders openly whether what are nominally American firms, but which now clearly operate as transnational enterprises, can be counted upon to act in ways that support the basic interests of the American public. He recognizes the decline in the authority and effectiveness of intermediate social institutions that once offered means for "political participation and negotiation–civic and charitable groups, political clubs, political parties, and trade unions." He calls for the building of new ties between government and business (with the stress, of course, on business) to handle the needs of public life. "Business enterprises . . . will largely replace geographic jurisdictions as conduits of government support for economic and human development. Companies, rather than state and local governments, will be the agents and intermediaries through which such assistance is provided."[7]

The gnawing question about proposals of this kind is why today's global corporations would want to lavish such attention upon any particular set of locals, Americans or otherwise? Reich recognizes that "at bottom the political issue is one of social membership. For whom does the business enterprise exist?" Much of the evidence he presents strongly suggests many of the major players in the private sector no longer have the best interests of the American people at heart. However, as a prudent advisor to would-be presidents of the United States, Reich is careful to sidestep the truly explosive issue at stake here, that of loyalty. Instead he issues the characteristic declaration of neoliberalism, calling for new kinds of cooperation between business, government and other segments of society to revitalize the national economy and build "human capital." "We are losing the competitive struggle because we cannot work together," he warns.[8]

But who, I wonder, is the "we" in such sentences? And "where" are they working to achieve their ends? Such proclamations, often heard in American politics these days, seem to call us back together as noble citizens of a Lockean society, one established on a traditional landscape. But what if some of those assumed to be committed to the existing social compact have, in large part, moved to a new place, *in* (electronic and corporate) *vacuis loci*?

The topic of loyalty has begun to surface in other corners of the business literature concerned with the new economic and technological horizons. In his fascinating book, *Tomorrow's Global Executive*, businessman Henry Ferguson argues that in fact all thriving firms must be transnational and that to be successful managers must wean themselves of any strong attachment to the culture and populace of any particular nation state. Executives must become global operators able to adapt quickly to the language, expectations and cultural practices of the people with whom they are dealing. "A global vision," he argues, "recognizes the interdependency that has gripped economic aspects of humanity in recent decades, especially since World War II. Will this vision ultimately lead to a new view of political identity? This is not clear today, though it is hard to see how the nationalistic identity of the past 200 years is a helpful construct in a world characterized by global markets. A global perspective and recognizing interdependency as a fact rather than a positive or a negative force may lead to decisions that appear hard-hearted or even 'disloyal.' Decisions that close plant doors in towns where the company grew up or lead to layoffs of loyal workers are always difficult. But such decisions appear differently when seen in a global perspective."[9]

The culture that matters now, in Ferguson's view, is corporate culture. To be successful in this dynamic, rapidly changing setting requires the cultivation of a new kind of personhood. "The global executive has to manage different scales of time, conduct research in unfamiliar territories, and plug into alien networks. The global executive has to master all of these challenges, then negotiate with the panache of a dancer, the verve of a gymnast, and the production skills of an impresario." He goes on to offer edifying advice on such matters as how to manipulate foreign business associates gracefully, when and how to offer bribes. Facing the challenges of transnational business, he notes, it is sometimes necessary to adapt one's morality to the circumstances at hand. "Morals are matters that the executive will have to deal with on a day-to-day basis, trying to strike a balance between what our culture has taught us are universal moral principles and the values upheld by the people he or she is dealing with."[10]

The revaluation of values discussed in the literature of transnational business mirrors a restructuring of relations of production and distribution taking place worldwide. In that process, what are first recognized as technical and economic necessities are later proclaimed as virtues. This is certainly true, for example, in the recent writings that link the need for "competitiveness" to the noble quest for something called "excellence," an excellence typically realized as the attaining of quality in one's product line, whatever that may be. What one sees in such cases are commonly held standards of personal morality and public good entering a process of thoroughgoing revision. The dimension of the new globalism that I have stressed here–its electronic, institutional dimension–is but one setting in which such changes are addressed. It is, however, a highly significant feature because without the new technologies of computing and telecommunications, such rapid and extensive restructuring of human relationships and understandings would not be possible. That is why these changes deserve to be noticed not only by those who study "communications" as a field of commercial and technological ventures, but by anyone who cares about the future of social and political institutions.

There are, of course, a great many questions such developments raise for both philosophy, social science and the general public. How will local communities respond to the power of transnational organizations and electronically embodied operations? How can one "hold responsible" persons and institutions whose precise location is ephemeral? How are such concepts as "rights," "duties," and "freedom," "justice," and "citizenship," to be applied to social entities created within electronic hardware and software?

Faced with these questions, it helps to notice that electronically embodied institutions themselves contain spaces of sorts, spaces defined by the architectures of their operating systems. Within these boundaries one finds structures of rules, roles, authorized practices, and relationships. The quality of membership one has within an electronic system is to a large extent a function of how that system has been designed. Entering this phenomenological domain, a person encounters social and political structures even though their features lack familiar, tangible forms.

In a typical case, scientific and technical professionals and higher level managers are greeted by a computerized version of the matrix form of organization in which there is at least superficial equality, an open access to information and one's co-workers. They experience computer networks as spheres of democratic or even anarchic activity. Especially for those ill at ease in the physical presence of others, the phenomenon of disembodied space seems to offer ecstatic fulfillment. Here are new opportunities for self-expression, creativity, and a sense of mastery. Some begin to feel they are most alive, most free when wandering through the network. Much of the contemporary popular literature on computers and society is absolutely giddy as it describes experiences of this kind. There are self-congratulatory proclamations that the collapse of hierarchy and liberation of humanity is at hand. As journalist Stewart Brand happily exclaims, "Information wants to be free."

Ordinary workers, on the other hand, often meet a much different set of possibilities. As they enter an electronically embodied office or factory, they are placed within a structure of top/down managerial control, required to take orders, complete finite tasks, and perform according to a set of standard productivity measures. What confronts them is a structure that incorporates the authoritarianism of the industrial workplace and augments its power in ingenious new ways. No longer are the time/motion measurements pioneered by Frederick Winslow Taylor limited by the awkward stopwatch carried from place to place by a wandering manager. Now workers' movements on the system can be ubiquitously and continuously monitored in units calculable to the nearest microsecond. For telephone operators handling calls, insurance company clerks processing claims, and keypunch operators doing data entry, rates of performance are recorded by a centralized computer and systematically compared to established standards. Failure to meet one's quota of phone calls, insurance claims or keystrokes is grounds for managerial reprimand or eventual firing. A report issued by the Office of Technology Assessment revealed that 4 to 6 million American office workers are already subject to such forms of computer-based surveillance.[11]

There are a great many other contexts in which social relationships and institutions are electronically embodied, for example those that either now or soon will shape possibilities in banking, credit, consumer spending, entertainment, and the like. For the time being, the building of such systems is regarded as the exclusive prerogative of private interests, those with the money to build and maintain them. Taking a broader view, however, it should be clear that rise of electronic places involves profoundly public questions. To an increasing extent the structure of such institutions defines the terms of our social membership, what rights and duties we have, what kinds of activities are permitted and which are not. Is the sum of such institutions to be a global village? A global democracy? A global bureaucracy? Global prison? Global asylum?

It is conceivable that we will soon find ourselves living among persons whose passports might well identify them as "digital transnationals." Perhaps some who read this already want to become such persons themselves. But before this occurs we owe it to ourselves to ponder the qualities of citizenship such digital transnationals will experience and the kinds of communities they will call home. It is fairly obvious how electronically embodied institutions are likely to affect productivity, profits, and possibilities for social control. All of that is taught in any good business school. But how, if at all, will these developments sustain the fundamental political ideals of the modern age: justice, equality, freedom, and democracy? How and where do such cherished ends enter the program?

The fact that electronically embodied institutions appear destined to undermine much of the authority of modern nation states is perhaps less worrisome than the fact that as they have arisen so far, these institutions lack any public philosophy whatsoever to guide the creation of their structures and operations. Within their boundaries it may be possible to build the conditions of an electronic *res publica*. But for the time being everyone views these institutions as if they were merely *res privata*. It is no longer in question whether or not there will be a politics for the ways we live in electronic space. The question is now: What complexion will that politics have?

Endnotes

[1]See Peter H. Lewis, "Harmony of Art and Science Lifts a Music Industry Barrier," *New York Times*, March 4, 1987, p. D6.

[2]For a description of some of the possibilities of computer teleconferencing see Kathleen Kelleher and Thomas B. Cross, *Teleconferencing: Linking People Together Electronically* (Englewood Cliffs, NJ: Prentice-Hall, 1985).

[3]For a description of Japanese initiatives in science and technology development see Sheridan Tatsuno, *The Technopolis Strategy: Japan, High Technology, and the Control of the Twenty-first Century* (New York, NY: Prentice-Hall Press, 1986).

[4]John Locke, *Two Treatises of Government*, a critical edition with an introduction and apparatus cirticus by Peter Laslett (New York: New American Library), pp. 330 and 393-394.

[5]Charles Norris Cochrane, *Christianity and Classical Culture; A Study of Thought and Action from Augustus to Augustine*. (New York: Oxford University Press, 1975), p. 213. For a discussion of the crisis in political thought during this period see Sheldon S. Wolin, *Politics and Vision: Continuity and Change in Western Political Thought* (Boston: Little, Brown and Company, 1960), chapter 3.

[6]See E. Lovelock, *Gaia: A New Look at Life on Earth* (New York: Oxford University Press, 1979); Bill Devall and George Sessions, *Deep Ecology* (Layton, UT: Gibbs M. Smith, 1984) and Michael Tobias, ed., *Deep Ecology*, (San Diego: Avant, 1984).

[7]Robert B. Reich, *The Next American Frontier* (New York: Times Books, 1983), p. 248.

[8]Ibid., pp. 267, 278 and 280.

[9]Henry Ferguson, *Tomorrow's Global Executive* (Homewood, IL: Dow Jones-Irwin, 1988), p. 41.

[10]Ibid., pp. 235-236.

[11]See *The Electronic Supervisor: New Technology, New Tensions*, OTA-CIT-333 (Washington, DC: U.S. Government Printing Office, September 1987).

NAZI BIOMEDICAL TECHNOLOGIES

Robert N. Proctor

Abstract: The place of science under National Socialism is different from what is commonly imagined. Science was not simply suppressed; certain kinds of science—especially biomedical science—actually flourished. Physicians were volunteers, and not just victims: doctors played an active part in helping to construct and administer the Nazi racial state, with its sterilization laws, its suppression of women's rights, its programs of systematic euthanasia and genocide. The Nazi racial synthesis was not acheived through technology alone, however: it took a powerful state to unleash the destructive potential of the biomedical community, turning racial theory into racial practice.

Philosophers have tended to ignore the experience of science under the Nazis, for several reasons. For one thing, discussions of science in philosophical literature are often couched in the abstract, as if there were some single, trans-historical logic of scientific method or scientific growth. Both Kuhn and Popper, for example, assume that there is only one way science grows: Kuhn describes a universal ontogeny of science, according to which scientific theories succeed one another through a well-defined and universal sequence of stages; Popper envisions a process of scientific growth through a community of scholars advancing bold conjectures, and testing these against experience (by persistent efforts to refute them).[1]

Lifeworld and Technology, eds. Timothy Casey and Lester Embree, Washington, D.C.: The Center for Advanced Research in Phenomenology and University Press of America, 1989.

Neither Kuhn nor Popper, however, pay particular attention to the political origins or impact of scientific ideas. For Kuhn, "world views" structure scientific inquiry–but nowhere does he explore how political values (or power relations) structure inquiry. For Popper, science is political, in the sense that ideals of freedom both in science and in society are not unrelated–but nowhere does he explore the role of ideology in scientific practice. (Indeed, in his view, ideologies are eliminated "in the long run" through continual criticism within the community of scholars.)[2]

Kuhn and Popper are interested in the logic of scientific ideas–how they arise, how they are accepted, how they are overthrown. If, however, we are interested not merely in the logic of theories, or the modes by which they change, but in the participation of knowledge in the production of power, we must focus on a different set of questions. Why do scientists support or resist the emergence of a particular political order? Why will scientists from one discipline (e.g., physics) support or resist a repressive regime to a different degree than another (e.g., anthropology)? Why do governments find it in their interest to support one kind of science and not another? And how are the theories or methods in a particular science shaped by the larger political order?

In liberal science theory, the assumption for most of this century has been either that science is *inherently democratic* (that is, it depends upon and contributes to democratic political formations), or that it is *apolitical*, and that the politicization of science implies its destruction. Science, in this view, must have only itself as a goal and a guide; science in the service of "interests" is no longer science. Implicit in this model is the view that science cannot flourish in a society that requires (asks?) science to serve interests other than itself. There is also a politics implicit in this judgment: namely, that science grows only on the soil of democracy, and that social forces which are hostile to democracy will be hostile to science. Such a view holds that the fate of science in a totalitarian regime is that it will be suppressed: the possibility that science (or scientists) might contribute to fascist movements is ruled out of court.

Given such a view, it is not hard to understand why philosophers have generally failed to come to grips with science under fascism: philosophers have generally assumed that there *could not have been* anything worthy of the name "science" under the Nazis. Indeed, as one philosopher in the

1960s put it, National Socialism "had hardly sufficient logical coherence to deserve the name of ideology."[3]

What I want to argue here, however, is that such a view is not borne out when we look at the experience of science under National Socialism. While it is true that certain fields, such as theoretical physics, did suffer–in most senses of that term–it is also true that others, such as anthropology or "genetic pathology," actually flourished–in a limited sense of that term.[4]

As we shall see, biomedical scientists played an active and leading role in the initiation, administration, and execution of each of the major Nazi racial programs. In this sense, science (and especially biomedical science) under the Nazis cannot simply be seen in terms of an essentially *passive and apolitical* scientific community responding to *external* political forces. This model of science as passive and apolitical in the face of Nazi racial politics underestimates the extent to which political initiatives arose from within the scientific community and the extent to which science and technology were integral parts of the Nazi program. It is a mistake, in other words, to view the relation of science and government in this period as essentially one of hostility: certain sciences were suppressed, but others did quite well.[5]

But I also want to argue that it is wrong to focus on science or technology *per se* as the guilty party in Nazi crime. The technologies used by Nazi ideologues were shaped by political priorities and nothing *inherent* in instrumental logic or the technological process brought about these crimes. Nazi science represents the triumph of a particular kind of science, but not of science in general. Ultimately a political movement–the seizure of state power by the Nazi party–that allowed forces hostile to life and liberty to be unleashed within the scientific community.

1. Biomedicine in the Nazi Racial State

The kind of science I want to focus on is what was known as "racial hygiene." At the end of the nineteenth century German social Darwinists, fearing a general "degeneration" of the human race, set about to establish a new kind of hygiene–a racial hygiene (*Rassenhygiene*)–that would turn the attention of physicians away from the individual or the environment, towards the human "germ plasm." In the eyes of its founders (Alfred Ploetz, Wilhelm Schallmayer), racial hygiene was supposed to complement personal and social hygiene; racial hygiene would provide long-run, preventive medicine for the German germ plasm by combating the disproportionate breeding of "inferiors," the celibacy of the upper classes, and the threat posed by feminists to the reproductive performance of the family.

In 1918, Hermann W. Siemens identified racial hygiene as "a political program that stands above all parties"; racial hygiene was based on "the facts of inheritance and variability." Inheritance taught that "man owes the essential part of his character, good or bad, physical or spiritual, to his genetic material (*Erbmasse*)." Variability taught that "men differ substantially in their genetic value (*Erbwert*)," and that "there are men who are genetically fit, and men who are genetically unfit." It followed, Siemens argued, that if those who are fit breed less than those who are inferior, the quality of the race will decline. The goal of a "positive racial hygiene" was to ensure that the fit leave more offspring than the unfit. If we fail in this, "all other earthly strivings will be in vain." More specifically, Germany will collapse in the face of an "Asiatic triumph."[6] (See figures 1 & 2)

Interestingly, the early racial hygiene movement was primarily nationalistic and meritocratic, more than it was anti-Semitic or Nordic supremacist. Eugenicists worried more about the indiscriminate use of birth control (by the "fit") and the provision of inexpensive medical care (to the "unfit") than about the breeding of superior with inferior races or many of the other themes we associate with the Nazis. Anti-Semitism played a relatively minor role in early racial hygiene. In fact, for Ploetz, father of the German movement (his 1895 treatise, in which he coined the term *Rassenhygiene*, served as the most important early book on this topic), Jews were to be classed along with the Nordics as one of the superior, "cultured" races of the world.[7]

By the mid-1920s, however, this had changed, and the right-wing faction of racial hygiene had merged with National Socialism. The conservative, anti-Semitic J. F. Lehmann Verlag took over publication of the *Archiv für Rassen- und Gesellschaftsbiologie* (the main racial hygiene journal) shortly after the war, and Nazi ideologues began to incorporate eugenics rhetoric into their discourse.

Figure 1) "Occupations and Number of Children." Graphic from Otto Helmut's *Volk in Gefahr* (Munich, 1937) depicts the disproportionate breeding of professionals and unskilled workers. The caption reads: "In 1912, for every marriage in Prussia there were the following numbers of children in the different professions: higher officials and independent professions, 2 children; government officials, 2.5 children; skilled workers, 2.9 children; unskilled workers, 4.1 children; land-laborers, 5.2 children."

Volksdeutsche Zukunft

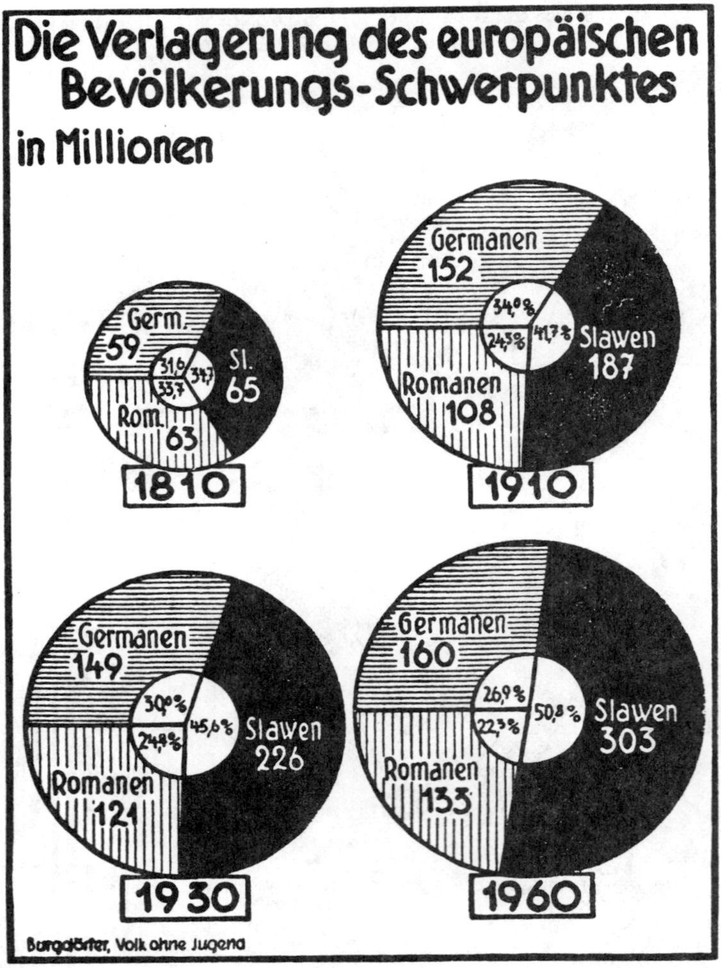

Figure 2) "The Displacement of European Peoples," depicting a steady decline of Germanic peoples in Europe. The chart forecasts that whereas in 1810, Germanic and Slavic peoples were roughly equal in numbers, by 1960, Germanic peoples will be outnumbered nearly two to one by Slavs. From Friedrich Burgdörfer, *Volk ohne Jugend*, 3rd ed. (Heidelberg, 1937), p. 387.

Traditionally, historians and philosophers have stressed the romantic origins of National Socialism: the links to German idealism, the nihilism of Nietzsche, or the Aryan supremacy of Gobineau, Chamberlain, and Lapouge. Biology (and especially *biologism*[8]), however, also played an important role in Nazi ideology. Fritz Lenz, for example (one of Germany's most prominent racial hygienists), praised Hitler in 1930 as "the first politician of truly great import who has taken racial hygiene as a serious element of state policy." Hitler himself was lauded as the "great doctor of the German people"; he once called his revolution "the final step in the overcoming of historicism and the recognition of purely biological values."

Biological imagery was important in Nazi literature in several ways. SS journals spoke of the need for "selection" to replace "counter-selection," borrowing their language directly from the social Darwinian rhetoric of the racial hygienists. Nazi leaders commonly referred to National Socialism as "applied biology"; indeed it was Lenz who originally coined this phrase in the 1931 edition of his widely read text-book on human genetics.[9]

The Nazi state was itself supposed to be *biologisch,* or organic, in two separate senses: in its suppression of dissent (the organic body does not tolerate one part battling with another) and in its emphasis upon "natural" modes of living. Nature and the natural mode of living was highly prized by Nazi philosophers: women were not supposed to wear make-up and legislation was enacted to protect endangered species. Hitler, as we know, was a vegetarian and did not smoke or drink, nor would he allow anyone to do so in his presence. One of the period's most powerful physicians (Hans Reiter) declared the Nazi state, judged by the organic views of its leaders, "first in the world."

Given the importance of biology in Nazi discourse, it is not surprising that *doctors* were among those most strongly attracted to the Nazi movement. In 1929, a number of physicians formed the "National Socialist Physicians' League" to coordinate Nazi medical policy and purify the German medical community of "Jewish Bolshevism." The organization was an immediate success, with nearly 3,000 doctors, representing six percent of the entire profession, joining the League by January 1933—that is, *before* the rise of Hitler to power. Doctors in fact joined the Nazi Party earlier and in greater numbers than any other professional group: by 1942, more than 38,000 doctors had joined the Nazi Party, representing about half of all doctors in the country. In 1937, doctors were represented in the SS seven times more often than was average for the employed male population; doctors assumed leading positions in German government and universities. (See figures 3-6)

2. Pawns or Pioneers?

One often hears that National Socialists distorted science, that doctors perhaps *cooperated* with the Nazi regime more than they should have, but that by 1933, as one emigré said, it was too late, and scientists had no alternative but to cooperate or flee. There is certainly some truth in this, but I think it misses the more important point that medical scientists were the ones who *invented* racial hygiene in the first place. Many of the twenty-odd university institutes for racial science and racial hygiene were established at German universities before the Nazi rise to power and it is fair to say that by 1932, racial hygiene had become a scientific orthodoxy in the German medical community. Racial hygiene was taught in the medical faculties of most German universities; racial hygiene was the subject of special lectures and seminars. The major expansion in this department occurs *before* Hitler comes to power; most of the 10-odd journals of racial hygiene, for example, were established long before the rise of National Socialism.

Racial hygiene was recognized as the primary research goal of two separate institutes of the prestigious Kaiser Wilhelm Gesellschaft: the Kaiser Wilhelm Institute for Anthropology in Berlin (1927-1945), directed by Eugen Fischer, and the Kaiser Wilhelm Institute for Genealogy in Munich (1919-1945), directed by the psychiatrist Ernst Rüdin. Both institutes performed research in the fields of criminal biology, genetic pathology, and what we today would call human genetics. Both also helped train SS physicians; both helped construct the "genetic registries" later used to round up Jews and Gypsies. Twin studies (that is, of identical twins raised apart) were among the leading preoccupations of these and other racial hygiene institutes; the purpose was to sort out the relative influence of nature and nurture in human character and institutions. This was to become one of the major priorities of Nazi medical research: in 1939, Interior Minister Wilhelm Frick ordered all twins born in the Reich to be registered with Public Health Offices for purposes of genetic research.

Figure 3) Cover page to *Deutsches Aerzteblatt*, official journal of the German Medical Association, July 1, 1933. The *Deutsches Aerzteblatt* was formed from a merger of Germany's two leading medical journals, the *Aerztliches Vereinsblatt* and the *Aerztliche Mitteilungen*, after the *Gleichschaltung* of their respective publishing societies.

Figure 4) Leonardo Conti, "Führer" of the German medical profession 1939-1943. From the *Deutsches Aerzteblatt*, 69 (1939): 323. Conti was head of the Berlin chapter of the National Socialist Physicians' League, and a leading member of the SS. His mother, "Nanna Conti," was head of the German Association of Midwives under the Nazis. Leonardo Conti succeeded Gerhard Wagner as "Reichsgesundheitsführer" in the spring of 1939, after Wagner's death; Conti helped plan the euthanasia operations, and was responsible for all health affairs of the Reich until Karl Brandt took over as Reich Health Kommissar in 1942. Conti committed suicide in a Nuremberg jail after the war, awaiting trial for war crimes. He was 45 years old.

Arischer Arzt

33 J., Kriegsteiln., vor der Zulassung, **übernimmt** gute **Allgemeinpraxis** in Stadt oder Land.
Angebote unter **W. 7552** befördert **Waibel & Co. Anzeigen-Gesellschaft, München 23,** Leopoldstr. 4.

Hilfsarzt

arisch, mit Aussicht auf baldige Einrückung in **Assistenzarzt-Stelle für neurologische Abteilung gesucht.** Nebeneinkommen aus Gutachtertätigkeit. Antritt **sofort.**
Krankenhaus Lankwitz, Berlin-Lankwitz, Viktoriastraße 51/59.

Figure 5) Advertisements for "Aryan physician" (*Arischer Arzt*) in the *Münchener Medizinische Wochenschrift*, Munich's leading medical weekly.

The Kaiser Wilhelm institutes for racial hygiene were funded very well: in 1935, the Rüdin Institute alone received 300,000 RM, more than the Kaiser Wilhelm Institutes for Physics and Chemistry combined. The largest such institution, however, was Otmar von Verschuer's Frankfurt Institute for Racial Hygiene. This institute had 67 rooms and several laboratories; this was where Josef Mengele did his doctoral research on the genetics of cleft palate, working under Verschuer. Mengele was also appointed assistant to Verschuer and provided "experimental materials" to the Institute (including eyes, blood, and other body parts) from Auschwitz as part of a study on the racial specificity of blood types (funded by the *Deutsche Forschungsgemeinschaft*). This, I should note, was one of the reasons blood groups were so actively studied in the 1930s: when Otto Reche founded the German Society for Blood Group Research in 1926, one of the main reasons he gave for this was to see if he could find a reliable means of distinguishing Aryans from Jews in the test tube.

Scientists, in other words, were not simply "pawns" in the hands of Nazi officials. But without a strong state to back it, racial hygiene was relatively impotent. It was not until 1933 and the triumph of Hitlerian fascism that the programs of the pre-Nazi period gained the support of officials willing to move aggressively in this area.

Die Ärzte der SS

Der Reichsführer SS und Chef der Deutschen Polizei stellt Bewerber für die Laufbahn des Arztes, Zahnarztes und Apothekers in die Waffen-SS und Polizei ein. Dem Arzt in der Waffen-SS und Polizei bietet sich neben dem rein truppenärztlichen Dienst ein weites Feld von Sonderaufgaben und Verwendungsmöglichkeiten. Er betreut als Arzt die Familien der ihm anvertrauten SS-Männer. Möglichkeiten zu klinischer und wissenschaftlicher Betätigung bieten sich ihm in den Lazaretten, Kur- und Genesungsheimen der Schutzstaffel. Für Forschungsarbeit stehen die SS-eigenen wissenschaftlichen Institute zur Verfügung. Bei der Lösung der großen bevölkerungspolitischen Aufgaben ist er in vorderster Front eingesetzt (Um- und Neusiedlung, Tuberkulosebekämpfung). Der Zahnarzt betreut als Truppenzahnarzt eine Einheit von Bataillonsstärke und überwacht ihre Familienangehörigen. Der SS-Apotheker arbeitet in Lazarettapotheken, Sanitätslagern und bei höheren Kommandostellen.

Figure 6) Ad for SS physicians, published in the *Aerzteblatt für Pommern*, official medical journal for Pomerania. The text announces that the Reichsführer of the SS and Chief of German Police (that is, Himmler) is looking for applicants for careers in medicine, dentistry, and pharmacy in the Waffen-SS. From the *Aerzteblatt für Pommern*, 8 (1941): 205.

3. The Sterilization Law

What were the practical results of Nazi racial hygiene? Three main programs–the Sterilization Law, the Nuremberg Laws, and the euthanasia operation–formed the heart of the Nazi program of medicalized, "racial cleansing." These (and especially the euthanasia program) were the programs that cleared the path for subsequent efforts to eliminate entire peoples from European soil.

On July 14, 1933, the Nazi government passed the Law for the Prevention of Genetically Diseased Offspring, or "Sterilization Law," allowing the forcible sterilization of anyone suffering from "genetically-determined" illnesses including feeble-mindedness, schizophrenia, manic depressive insanity, genetic epilepsy, Huntington's Chorea, genetic blindness, deafness, or "severe alcoholism." The measure was drawn up after a series of meetings by several of Germany's leading racial hygienists, including Fritz Lenz, Alfred Ploetz, Ernst Rüdin, Heinrich Himmler (who had been active in breeding chickens prior to 1933), Gerhard Wagner, "Führer" of the National Socialist Physicians' League, and Fritz Thyssen, the industrialist.

In 1934, 181 Genetic Health Courts and "Appellate Genetic Health Courts" were established throughout Germany to adjudicate the Sterilization Law. The courts were usually attached to local civil courts and presided over by two doctors and a lawyer, one of whom had to be an expert on "genetic pathology." According to the provisions of the law, doctors were required to register every case of genetic illness known to them; physicians could be fined 150 RM for failing to register any genetic defective. Physicians were also required to undergo training in "genetic pathology" at one of the numerous racial institutes established throughout the country. Also in 1934, the German Medical Association founded *Der Erbarzt* (The genetic doctor) to help physicians determine who should be sterilized; the journal also described the latest sterilization techniques.

Estimates of the total number of people sterilized in Germany range from 350,000 to 400,000. In some areas, more than one percent of the total population was sterilized. Compared with the demands of some racial hygienists this was relatively modest: Lenz, for example, had argued that as many as 10-15 percent of the entire population were defective and ought to be sterilized.

As a consequence of the law, sterilization research and engineering rapidly became one of the largest medical industries. Medical supply companies (such as Schering) designed sterilization equipment; medical students wrote more than 180 doctoral theses exploring the criteria, methods, and consequences of sterilization. There were obvious incentives for developing more rapid sterilization techniques, especially for women, for whom the standard procedure of tying the fallopian tubes could involve a hospital stay of more than a week. The most important of these techniques was the development of "operationless" sterilization involving scarification of the fallopian tubes through injections of super-cooled carbon dioxide. In 1943, the gynecologist Clauberg announced to Himmler that, using such a technique and with a staff of ten men, he could sterilize as many as 1,000 women per day. Scarification was not the only new technique developed: according to one study, 12 percent of all sterilizations were performed with x-rays, a technique also used in the United States at this time.

I should also mention that it was the United States that provided the most important model for German sterilization laws.[10] By the late 1920's 15,000 individuals had been sterilized in the U.S.–most while incarcerated in prisons or homes for the mentally ill. Nearly half of these took place in California–most of the others in northern states (not, as one might have expected, in the south). German racial hygienists throughout the Weimar period expressed their envy of American achievements in this field, warning that unless Germany made progress in this area, America would become the world's racial leader.

4. Racial Medicine: The Control of Women

Racial domination and the elimination of the weak and unproductive were not the only forms of oppression in the Nazi regime. One aspect of Nazi ideology that has come under increasing scrutiny in recent years is the masculine and *machismo* nature of that ideology. Of the two-and-a-half million members of the Nazi Party in 1935, only five-and-a-half percent were women; women were explicitly barred from top positions in the Nazi ranks as early as January, 1921. Nazi medical philosophers were quite explicit about their feelings on this matter: a 1933 editorial of the journal of

Kinderwagenparade vor der Beratungsstelle

Figure 7) Baby carriages lined up outside one of Germany's "Advisory Centers for Genetic and Racial Care" (*Beratungsstelle für Erb- und Rassenpflege*). From the *Aerzteblatt für Berlin und Kurmark*, 43 (1938): 352.

the Nazi Physicians' League announced that the National Socialist movement was "the most *masculine* movement to have appeared in centuries."

One of the initial thrusts of Nazi policy was to take women out of the workplace and return them to the home, where they were to have as many children as possible. Fritz Lenz, for example, argued that any woman with fewer than 15 babies (!) by menopause should be considered "pathological." The government was more modest, pushing what it called the *four-child family* ideal. On December 16, 1938, Hitler announced the establishment of the "Honor Cross of German Motherhood" (modeled on the Iron Cross): awarded in bronze for four children, silver for six, and gold for eight.[11] After 1938, all public officials (including professors!) were required to marry or else resign; medical journals published the names of unmarried or childless colleagues.

Nazi population policy, directed towards what Interior Minister Frick called "the solution to the woman question" was remarkably successful: the birth rate jumped from 14.7/1000 in 1933 to 18/1000 in 1934, representing what Friedrich Burgdörfer called an unprecedented achievement in world population history, and a victory in the "war of births." (See figure 7).

Yet the German "victory" was one that set efforts at women's liberation back nearly a hundred years. According to official Nazi policy, unmarried women were not considered citizens but were relegated instead to the status of *Staatsangehöriger* (deprived, that is, of certain rights), to which Jews were also initially assigned. As of 1936, women were banned from becoming lawyers, judges, and a host of other professions; quotas on the proportion of female students at German universities were reinstated.

At the same time that forced sterilization and abortion were instituted for individuals of inferior genetic stock, sterilization and abortion for "healthy" German women were declared illegal, and punishable (in some cases by death) as a "crime against the German body." (Bavaria's official medical journal declared abortion a form of "treason.") Access to birth control in all forms was also severely curtailed—except of course for Jews, for whom birth control was generally available, and encouraged.

It is not well known, but there is one final aspect of gender that has escaped most discussions of the holocaust. In 1939 there were substantially more Jewish women in Germany than Jewish men; indeed the ratio was roughly fourteen women for every ten men[12] (men had presumably managed to emigrate in greater numbers). Historians have largely ignored the fact that among German Jews killed in concentration camps (including most of the 167,000 Jews living in Germany in 1941), some 60 percent were women. Women also outnumbered men among the Gypsies killed at Auschwitz.[13]

5. The Nuremberg Laws

In the fall of 1935, Hitler signed into law the so-called "Nuremberg Laws" excluding Jews from citizenship and preventing marriage or sexual relations between Jews and non-Jews. A further measure, the "Marital Health Laws," required couples to submit to medical examination before marriage to see if "racial pollution" might be involved.

I cannot go into detail here on the operation of these laws; the story has been told elsewhere.[14] Let me simply point out that the Nuremberg Laws were considered public health measures and were administered primarily by physicians. In early 1936, when the Marital Health Laws went into effect, responsibility for administering the laws fell to the marital counseling centers established in the Weimar period. The centers were greatly expanded beginning in 1935; they were renamed "genetic counseling centers" and attached to local public health offices. The Nuremberg Laws, along with the Sterilization Law, were two of the primary reasons expenditures and personnel for public health actually *expanded* under the Nazis.

I should also note that, as with the Sterilization Law, here, too, German racial theorists learned from the Americans. In 19312, Bavarian Health Inspector Walter Schultze noted that the U.S. must be regarded as a nation where "racial policy and thinking has become much more popular than in other countries." Schultze was referring not just to sterilization laws, but to American immigration restriction laws—especially the 1924 Immigration Restriction Act, according to which annual immigration into the U.S. was cut by some 95 percent.

In subsequent years, racial hygienists looked to other aspects of American racial policy for instruction. Nazi physicians on more than one occasion argued that German racial policies were relatively "liberal" compared with the ways blacks were treated in the United States. Evidence for this was usually taken from the fact that in several southern states, a

Figure 8) Illustration from the popular magazine *Neues Volk* depicting restrictions on the civil rights of blacks in the United States. Caption at top reads "How Racial Questions Arise: White and Black in America." Map indicates states where:

- Blacks have no voting rights, are not allowed to marry whites, and are segregated from whites (dark black).
- Blacks face all of the above restrictions with the exception of voting rights (darkly spotted).
- Black-white intermarriage is not allowed; blacks and whites attend separate schools (dark cross-hatched lines).
- Black-white intermarriage is the only restriction blacks face (light cross-hatched lines).

Nazi journals used the example of racial legislation in the US to defend the suppression of civil rights of Jews in Germany. From *Neues Volk*, March 1, 1936, p. 9.

person with 1/32 black ancestry was legally black, whereas if someone was 1/8 Jewish in Germany (and for many purposes, 1/4 Jewish), that person was legally "Aryan." Nazi physicians spent a great deal of time discussing American miscegenation legislation; German medical journals reproduced charts showing the states in which blacks could or could not marry whites, could or could not vote, and so forth. (See figure 8).

In 1939, Germany's leading racial hygiene journal reported that the University of Missouri had refused to admit black students. The same year, the journal reported the recent refusal of the American Medical Association to admit black physicians to its membership. German physicians had only recently (in 1938) barred Jews from practicing medicine (except on other Jews); racial theorists were thereby able to argue that Germany was "not alone" in its efforts to preserve racial purity.[15]

6. Euthanasia

In early October, 1939, Hitler issued orders that certain doctors be commissioned to grant "a mercy death (*Gnadentod*) to patients judged incurably sick by medical examination." By August 1941, when the so-called "euthanasia" operation was largely brought to a close (see below, however), more than 70,000 patients from German mental hospitals had been killed in an operation which provided the stage rehearsal for the subsequent destruction of the Jews, homosexuals, Communists, Gypsies, Slavs and prisoners of war.

The idea of the destruction of "lives not worth living" did not begin with the Nazis, but had been discussed in legal and medical literature long before the Nazi rise to power. And not just in Germany. In 1935, the same year Egas Moniz invented the lobotomy, the French-American Nobel Prize winner Alexis Carrel suggested in his book, *Man the Unknown*, that the criminal and mentally ill should be "humanely and economically disposed of in small euthanasia institutions supplied with proper gases." Six years later, as German psychiatrists were sending the last of their patients into the gas chambers, an article appeared in the *Journal of the American Psychiatric Association* calling for the killing of retarded children.

The fundamental argument for forcible euthanasia was economic: euthanasia was justified as a kind of "preemptive triage" to free up beds. This became especially important in war time. I want to stress this: things can happen in war that would not be tolerated in peacetime. And in fact, the euthanasia operation was consciously timed to coincide with the invasion of Poland. The first gassings of mental patients occurred at Posen, in Poland, on October 15, 1939, just 45 days after the invasion of that country marking the beginning of the Second World War.[16] In Germany itself, after the end of the first phase of euthanasia in August 1941, euthanasia became part of normal hospital routine: handicapped infants were regularly put to death; persons requiring long-term psychiatric care and judged "incurable" suffered a similar fate. The war-economy claimed many victims: after bombing attacks made space short, psychiatric institutes were cleared out (their patients murdered) to make up for lost space. In at least one instance, a home for the elderly was emptied in this fashion–its inhabitants sent to the concentration camp at Meseritz-Obrawalde (in Poland), where they were killed.[17]

The importance of war can also be seen in the fact that during the First World War, half of all German mental patients starved to death (45,000 in Prussia alone, according to one estimate)–they were simply too low on the list to receive rations. In the Nazi period, the starvation of the mentally ill, the homeless, and other "useless eaters" became official state policy, after a prolonged propaganda campaign to stigmatize the mentally ill and handicapped as "lives not worth living." (See figure 9).

One should recall that the euthanasia program was planned and administered by leading figures in the German medical community. When the first experiments to test gasses for killings took place at Brandenburg Hospital in January 1940, Viktor Brack, head of the operation, emphasized that such gassings "should be carried out only by physicians." Brack cited the motto: "the needle belongs in the hand of the doctor."

It is also important to appreciate both the *banality* and the *popularity* of the euthanasia operation. In 1941, for example, the psychiatric institution at Hadamar celebrated the cremation of its ten-thousandth patient in a special ceremony, where everyone in attendance–secretaries, nurses, and psychiatrists–received a bottle of beer for the occasion. The operation was also popular *outside* the medical community. Parents were made to feel shame and embarrassment at having to raise an abnormal or malformed child: hospital archives are filled with letters from parents writing to health authorities requesting their children be granted euthanasia.

Historians exploring the origins of the Nazi destruction of "lives not worth living" have only in recent years begun to stress the links between the euthanasia operation on the one hand and the "final solution" on the other. And yet the two programs were linked in both theory and in practice. The most important theoretical link was what might be called the "medicalization of anti-Semitism," part of a broader effort to reduce a host of social problems–unemployment, homosexuality, crime, "antisocial behavior," etc.–to medical or ideally *surgical problems.*

The "Jewish question," for example was commonly cast as a public health problem. One of the leading research priorities in this period was to study the racial specificity of disease (I mentioned earlier the case of blood group research). In 1935, when Gerhard Wagner addressed the Nuremberg Party Congress, he was able to cite extensive medical research proving that various "diseases," including homosexuality, insanity, feeble-mindedness,

Figure 9) "Genetically Diseased" (*Erbkrank*). This was a picture commonly reproduced in Nazi medical and paramedical journals in order to stir up sentiments against so-called "lives not worth living" (*lebensunwerten Lebens*). This picture is from *Ziel und Weg*, 5 (1935): 466.

hysteria, suicide, gallstones, bladder and kidney stones, neuralgia, chronic rheumatism, and flat feet, were more common among Jews than non-Jews. In the same speech at which he proposed the destruction of the mentally ill, Wagner declared that "Judaism is disease incarnate."

The medicalization of deviance continued into the war years. The first Nazi-administered Jewish ghettos in Poland, for example, were justified in terms of a "quarantine"; permission to enter or leave the ghettos was granted only by medical authorities. This of course eventually also became a self-fulfilling prophecy: medical journals in the early 1940s denounced the Jewish "race" as a diseased race and were able to invoke statistics on typhus levels in the Warsaw ghetto to prove this.[18]

In the course of the late 1930s, German scientists proposed a number of different solutions to the "Jewish question." The agronomist Dr. Hefelmann suggested exporting all Jews to Madagascar. Dr. Philip Bouhler proposed sterilizing all Jews by x-rays. Viktor Brack recommended sterilization of the two to three million Jews capable of work, who might be put to use in Germany's factories. German medical authorities also devoted themselves to this problem: during the early war years, *Das Deutsche Aerzteblatt*, the official journal of the German Medical Association, published a regular column on "Solving the Jewish Question." (See figure 10).

The ultimate decision to gas the Jews emerged from the fact that the technical apparatus already existed for the destruction of the mentally ill. In the fall of 1941, with the completion of the bulk of the euthanasia operation, the gas chambers at psychiatric hospitals were dismantled and shipped East, where they were reinstalled at Majdanek, Auschwitz, and Treblinka. The same doctors, technicians and nurses often followed the equipment. In this sense, there was a continuity in both theory and practice between the destruction of the "lives not worth living" in Germany's mental hospitals and the destruction of Germany's ethnic and social minorities.

Zur Lösung der Judenfrage

Berichte aus dem Ausland

Frankreich:

Auf Grund des Gesetzes vom 16. August 1940 sind die Reihen der Ärzteschaft weiter von Juden gesäubert worden. Für das Gebiet Frankreich, Nordafrika und für die Kolonien ist eine erste Liste aufgestellt worden, die nicht weniger als 6000 jüdische Ärzte umfaßt.

Nach einer Aufstellung des Generalsekretariates für öffentliche Gesundheit beträgt die Zahl der ausländischen Ärzte, die unter das Gesetz fallen, 5000, die mit ganz wenigen Ausnahmen Juden sind.

In zehn Jahren, von 1929 bis 1939, war der Hundertsatz der jüdischen Ärzte in Frankreich von 17 auf 35 gestiegen. Auch viele andere Berufe waren von Juden förmlich überschwemmt worden.

Finnland:

Die Juden in Finnland haben es bisher vermieden, in größerer Zahl politisch oder wirtschaftlich sichtbare Stellungen einzunehmen. Ihr indirekter Einfluß ist nichtsdestoweniger bedeutend, zumal sich ein erheblicher Teil des Mobilienkapitals in ihren Händen befindet und der jüdische Anteil an einzelnen Geschäftszweigen, wie zum Beispiel in der Konfektion und im Pelzwarenhandel, ständig zunimmt. Die Tageszeitung der vaterländischen Volksbewegung in Finnland, „Ajan Suunta", befaßt sich in längeren Ausführungen mit der Judenfrage in Finnland und betont, daß hier bisher noch nicht einmal die ersten Schritte zur Lösung dieses wichtigen Problems getan seien. Die völkische Erneuerung könne aber auch in Finnland erst ihr Werk tun, wenn man an die Fragen herangehe, die die Stellung des Judentums betreffen.

Irak:

Die irakische Regierung hat neue verschärfte Maßnahmen gegen die Wucherer und Kriegsgewinnler ergriffen, nachdem es jüdische Exporteure in Bagdad fertiggebracht haben, die Preise für gewisse Einfuhr-Fertigwaren in einer Woche um das Zwei- bis Dreifache hinaufzuschrauben.

Figure 10) "Solving the Jewish Question," a regular column appearing in the *Deutsches Aerzteblatt*, official journal of the German Medical Association. This column is from the 1942 volume of the journal, p. 71.

8. Conclusion

Most leading German physicians supported the Nazis. Why? Physicians commonly boasted that their profession had shown its allegiance earlier and in greater strength than any other group. But why?

First of all, we should recall that the medical profession at this time was quite conservative. Prior to 1933, the leadership of the profession was dominated by the *Deutschnationalen*—a German nationalist party that subsequently threw its support to Hitler. Not all physicians, of course, were conservative. The profession was politicized and polarized after the economic collapse in the late 1920s and early 1930s; physicians moved from the center to the left or the right. The socialists (and communists), however, were always a minority in the German medical community. By the end of 1932, the Nazi Physicians' League was twice as large as the Association of Socialist Physicians (3,000 vs. 1,500 members). In the Reichstag elections leading to the Nazi seizure of power, *nine* physicians were elected to represent the Nazi Party; only *one* physician was elected to represent the socialists.

Apart from this conservatism, it is possible to argue that there was a certain *ideological affinity* between medicine and Nazism at this time. Many physicians were attracted by the importance given to race in the Nazi view of the world. Race at this time was generally considered a medical or anthropological concept (most anthropologists were trained as physicians); physicians were intrigued by the Nazi effort to *biologize* or *medicalize* a broad range of social problems, including crime, homosexuality, the falling birth rate, the collapse of German imperial strength, and the "Jewish and Gypsy problems."

The Nazis, in turn, were able to exploit both the intimacy and the authority of the traditional physician-patient relationship. Crudely put: you could do things with doctors that would have been much harder without. Doctors served as executioners on German submarines; they routinely performed "selections" (of people to be killed) in concentration camps.[19] Himmler recognized the special role of physicians in this regard: on March 9, 1943, the Reichsführer of the SS issued an order that henceforth only physicians trained in anthropology could perform selections at concentration camps. Medicine also served as a *disguise*: in Buchenwald 7,000 Russian prisoners of war were executed in the course of supposed "medical exams," using a device disguised as an instrument to measure height. (See figure 11.)

There is a further element. The rise of the Nazis coincides with a period of concern about what was widely known as "the crisis" in modern science and in medicine: a crisis associated with the increasing specialization and bureaucratization of science, a crisis traced alternatively to urbanization, capitalism, Bolshevism, materialism, or any of a host of other real and/or apparent threats to human health and well-being. The Nazis promised to restore Germany to a more natural (*biologische*) way of living: Nazi philosophers rejected separate sciences "for the laboratory and the bedside"; science in the new state would challenge Galileo's claim "to measure what is measurable, and make measurable what is not." Nazi philosophers promised a future with "more Goethe, and less Newton."

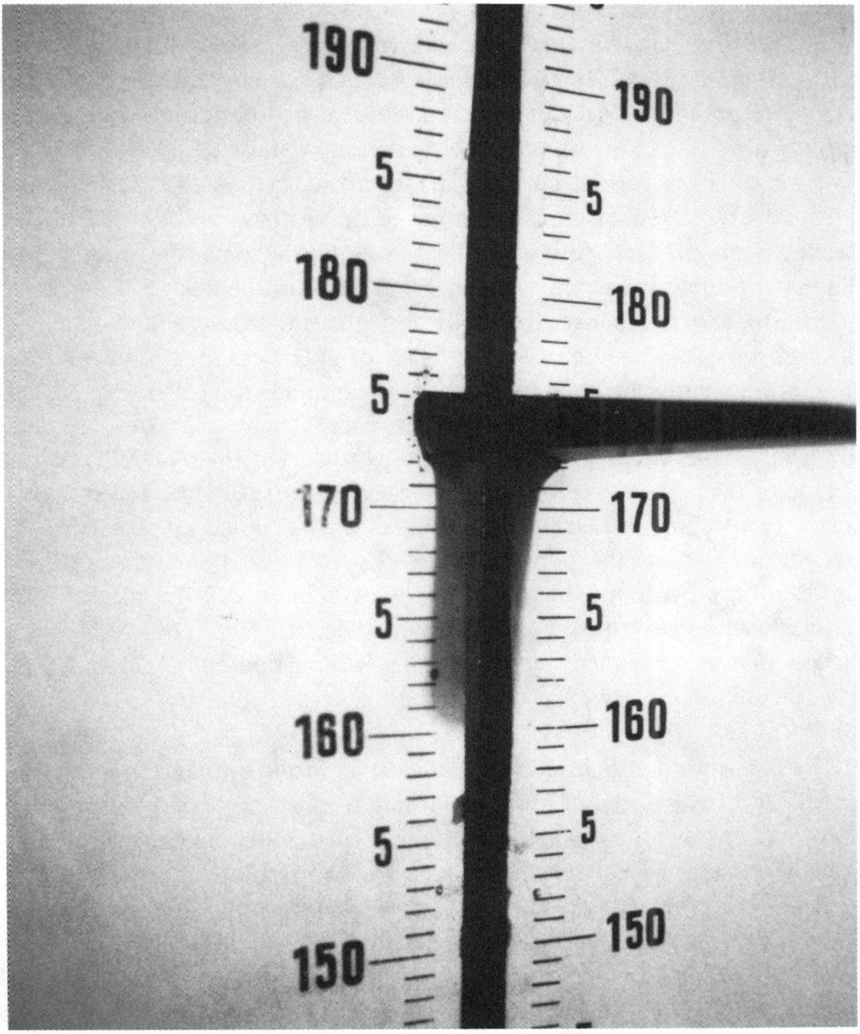

Figure 11) Murder disguised as medicine. This was a device used to kill 7,000 Russian prisoners of war in Buchenwald. Prisoners were executed in the course of supposed "medical exams"; after stepping up to have their height measured, prisoners were shot in the neck by a gun hidden behind this device.

In such a climate, Jews became a convenient scapegoat for all that was wrong in modern medicine. This was especially easy because Jews were in fact quite prominent in the German medical profession: 60 percent of Berlin's physicians, for example, were either Jewish or of Jewish ancestry. Jews were blamed for the bureaucratization and scientization of medicine; Jews were accused of transforming medicine from a "calling" into a "trade," from an "art" into a "business." National socialism promised to eliminate the Jews, restoring the honor and status of the doctor.

And in a certain sense, the medical profession might even be said to have *prospered* under the Nazis. In terms of both personnel and funding, the German medical community was enlarged under the Nazis–despite the banishing of the Jews and the communists. It may even be true that physicians achieved a higher status in the Nazi period than any time before or since. During the twelve years of Nazi rule, for example, the office of *Rektor* (president) at German universities was occupied by physicians 30 percent of the time; this contrasts with 19 percent for the decade prior to the rise of the Nazis and 18 percent for the two decades following the Nazi period. Doctors also prospered financially under the Nazis. In 1926 lawyers earned an average annual salary of 18,000 RM, compared with only 12,000 RM for physicians. By 1936 doctors had reversed this, and now earned 2,000 RM more than lawyers.

The Nazis were, it is true, hostile to certain forms of theoretical science, especially certain forms of physics and mathematics. But they did support many areas of the sciences, especially the applied and engineering sciences and social sciences such as psychology and anthropology.[20] In the case of medicine, it is interesting that there is no massive exodus from the editorial boards of medical journals in the period 1932-35. Medical journals did *not* close their doors. If you go to the New York Academy of Medicine, or Stanford's Lane Library, or any other major medical library, you can find more than 150 German medical journals published continuously through the Nazi period–more than 100 meters of shelf space of journals! Few medical journals ceased publication in the early years of the regime; most of those which did, did not close until the mid-war years.[21] In fact, some 30-odd *new* medical journals begin publishing during the Nazi period. Several of these are still published today.

Biomedical science was *not*, in other words, simply destroyed by the

Nazis—the story is more complex. The Nazis suppressed some areas, and encouraged others. The Nazis supported extensive research into ecology, public health, carcinogenesis, human genetics, criminal biology, and (of course) racial and sociobiology.[22] The Nazi government funded elaborate research on the effects of exposure to x-rays and heavy metals; some of the first reliable studies of the health effects of asbestos were done in this period. The Nazis were the first (so far as I'm aware) to initiate bans on smoking in public buildings;[23] Nazi leaders organized massive support for midwifery, homeopathy, and a number of other areas of heterodox medicine. Nazi physicians recognized the importance of a diet high in fruit and fiber; the Nazis passed a law requiring every German bakery to produce whole-grain bread. Nazi physicians restricted the use of DDT, and denied pregnant women tobacco rationing coupons on the grounds that nicotine could harm the fetus. Nazi physicians stressed the importance of preventive rather than curative medicine; racial hygiene was supposed to provide "long-run" preventive care for the German germ plasm, complementing shorter term social and individual hygiene.

I've stressed the continuity with pre-1933 traditions; I do not have space here to discuss the important post-war continuities. Let me simply note in conclusion four points.

First: it is important to appreciate not just the extent to which the Nazis were able to draw upon the imagery and authority of science, but also the extent to which Nazi ideology shaped the practice of science.

Second: scientists were not bystanders, nor even pawns: many (not all, but not a few) helped to construct the racial policies of the Nazi state. It is probably as fair to say that Nazi racial policy emerged from *within* the scientific community as to say that it was imposed *upon* the scientific community.

Third: it is commonly said that the Nazis "politicized" science and that much of what went wrong under the Nazis can be traced to this politicization. The argument I have made here is that one cannot consider the experience of the medical profession in terms of a simple "use and abuse" model of science: among physicians, there were as many volunteers as victims; no one had to *force* physicians to support the regime. Hefelmann testified to this effect in the euthanasia trial at Limburg in 1964: "no doctor was ever ordered to participate in the euthanasia program; they came of their own volition."[24]

The Nazis did not have to politicize science; in fact, it is probably fair to say that the Nazis "de-politicized science"—in the sense that they *removed* from science the political diversity that makes the politics of science (or medicine, or technology) interesting. The Nazis destroyed the vigorous socialist medical tradition that had transformed Weimar medical practice (establishing local outpatient clinics, self-help networks, etc.). The Nazis denounced the "political" as a category of discussion: the German state was to be a *Volksstaat*, not a *Parteistaat*; National Socialism was to be counted a "movement," not a "party." The Nazis sought to transform problems of racial, sexual, or social deviance into "medical problems"; Germany's social and political problems would be solved by "cleansing," "disinfection," "surgery." Murder was practiced in the name of quarantine, apartheid in the name of public health.

Finally: I do not want to leave the impression that the horrors of this period can be attributed to anything *inherent* in science or in medicine, or even in "technocracy" or the rule of technical elites. It took a powerful state to concentrate and unleash the destructive forces within German medicine and without that state, science would have remained impotent in this sphere. In the midst of a war engineered by an aggressive, expansionistic state, Nazi ideologues were able to turn to doctors and scientists to carry out programs that even today stand as unexcelled exemplars of evil. The collaboration of scientists made this disturbingly easy, frightfully free of hitches.

Where was the resistance, and why was it so feeble? Why were so many so willing to serve? The myth of "following orders," invented for post-war tribunals, has never been convincing. The questions remain.

Endnotes

[1]This essay represents a synopsis of certain arguments developed at length in my *Racial Hygiene: Medicine Under the Nazis* (Harvard University Press, 1988); further sources and documentation for the questions raised in this essay can be found in that work.

[2]See, for example, Popper's "Twelfth Thesis" in Theodor W. Adorno et al., *The Positivist Dispute in German Sociology* (New York, 1976), p. 96.

[3]The *Encyclopedia of Philosophy* (8 Vols., New York, 1967) suggests that unlike Marxism, fascism and Nazism "were not coherent doctrines, and their proponents never made more than a pretense of reconciling theory and practice" (vol. 6, p. 385). Another article states that National Socialism "had no doctrine in the proper sense of the word. It was fundamentally hostile to philosophy, and the ideas of orthodoxy and heresy remained foreign to it . . . The ideas–few in number, and crude as well as shallow–that formed the hard core of the national socialist creed had hardly sufficient logical coherence to deserve the name of ideology . . . National Socialism was a perversion of philosophy rather than a philosophy" (vol. 3, pp. 309-310).

[4]I use the term "flourish" in a qualified and deliberately provocative sense. It is important to distinguish not only between different sciences in this regard (e.g., between human genetics and theoretical physics), but also between different senses of the term "flourish." I am not claiming that *all* sciences flourished, nor that science flourished in a moral or theoretical sense. I mean that science (and only certain sciences, and only for a time) flourished in a professional or purely *quantitative* sense–numbers of journal articles published, research expenditures, etc. One cannot doubt that science as a whole suffered profoundly in both a *moral* and *theoretical* sense during this period: morally, science is put in the service of oppressive injustice and ultimately genocide; theoretically, the position of world leadership occupied by German science prior to 1933 is largely destroyed by the expulsion of the Jews and the reign of terror imposed by Nazi rule.

The collapse of Germany during the war dealt a further blow to German scientific institutions from which neither of the two modern Germanies has fully recovered.

[5]Interestingly, those areas of science which were suppressed by the Nazis–relativity theory, for example, or quantum mechanics–have been those which have, until recently, received the most attention from historians and sociologists. This is probably due to the fact that these fit the traditional image of the fate of science under totalitarian rule–namely, that science and totalitarian rule are mutually incompatible.

[6]Hermann W. Siemens, "Was ist Rassenhygiene?" *Archiv für Rassen- und Gesellschaftsbiologie,* 12 (1916/1918): 281-282.

[7]Alfred Ploetz, *Die Tüchtigkeit unsrer Rasse und der Schutz der Schwachen* (Berlin, 1895), p. 138.

[8]Biologism (or biological determinism) is the view that "nature" is more important than "nurture" in the formation of human character and institutions. Typically it is the view that men and women, or different races, have different inborn talents or disabilities–that certain individuals, for example, are "born criminals," or that the occupations people choose are largely determined by their inherited capacities. See Richard Lewontin et al., *Not in Our Genes* (New York, 1984).

[9]Fritz Lenz, *Menschliche Auslese und Rassenhygiene (Eugenik)*, 3rd ed. (Munich, 1931), p. 417.

[10]The fact that Americans had pioneered forcible sterilization was important in post-war deliberations concerning which among the Nazi racial programs should be considered "war crimes." Nuremberg authorities generally did not consider forcible sterilization criminal. After the war, individuals who had been sterilized received compensation as victims of Nazi oppression only if they could prove they had been sterilized *illegally*, that is, only if they could prove they were *not* in fact genetically feeble-minded, alcoholic, epileptic, etc. The reason for this was obvious–that similar programs had been widespread in the U.S.

[11]Similar awards were not uncommon in other countries at this time. In 1926, a Toronto millionaire willed $700,000 to the Canadian woman who, by the year 1936, would have the most children. I don't know who won, but as of 1930 there was a certain Mrs. Brown who, at the age of 40, had already given birth to 26 children!

[12]The May 15, 1939 census showed there to be 380,892 Jews in the *Altreich* (Germany prior to the annexation of Austria), down from 722,000 in 1933. The census showed a ratio of women to men of 1,366 to 1,000. See the *Archiv für Rassen- und Gesellschaftsbiologie*, 34 (1940-1941): 313.

[13]See Gisela Bock, *Zwangssterilisation im Nationalsozialismus* (Opladen, 1986), p. 13.

[14]See my *Racial Hygiene: Medicine Under the Nazis* (Cambridge, Mass.), esp. chapter 6.

[15]See "Keine Negerärzte in der amerikanischen Standesorganisation," *Archiv für Rassen- und Gesellschaftsbiologie*, 33 (1939-1940): 276; also 33 (1939-1940): 96. In 1930 there were some 5,000 black American physicians among a population of some 13,000,000 blacks. The appeal by America's black physicians was placed at the annual meeting of the AMA in St. Louis; the appeal was rejected. AMA officials claimed that the problem of black physicians was a problem only in Southern states, and that this problem should be solved in those states, not by a national body such as the AMA (*ibid*). America's leading medical organization remained essentially a segregated body until the mid-1950s.

[16]Institut für Zeitgeschichte, *Medizin im Nationalsozialismus* (Munich, 1988), pp. 24-25.

[17]*Ibid.*, pp. 25-26.

[18]After the war, the Nazi chief of police for occupied Warsaw (Arpad Wigand) defended shooting several thousand Jews attempting to leave the ghetto on the grounds that the area had been quarantined.

[19]Benno Müller-Hill, *Murderous Science: Elimination by Scientific Selection of Jews, Gypsies, and Others, Germany 1933-1945* (Oxford, 1988).

[20]For the case of anthropology, see my "From *Anthropologie to Rassenkunde*: Concepts of Race in German Physical Anthropology," in *Bones, Bodies, Behavior: Essays on Biological Anthropology*, ed. George Stocking (Madison, 1988).

[21]Many of the journals closed by the Nazi regime were associated with either communist or socialist causes. Journals primarily devoted to the economic affairs of the profession were also reorganized and began publishing once again only after the *Gleichschaltung* of the profession.

[22]One of the earliest uses of the term "sociobiology" in English can be found in the Nazi criminologist Hans von Hentig's book, translated into English as *The Criminal and his Victim, Studies in the Sociobiology of Crime* (New Haven, 1948).

[23]In 1939, the city of Dresden enacted a law prohibiting smoking in all public buildings. See the *Archiv für Rassen- und Gesellschaftsbiologie*, 33 (1939-1940): 274.

[24]Fredric Wertham, *A Sign for Cain* (New York, 1966), p. 167.

TECHNOLOGY AND LIBERATION

Algis Mickunas

Abstract: While technology purports to be instrumental, it is based on a specific ontological intentionality. The latter structures political and scientific enlightenments, each claiming the possibility of liberation. Political enlightenment claims that liberation requires democratic institutions, while scientific enlightenment postpones liberation until the experts have established the proper conditions for freedom. The tension and promise of modernity is based on the efforts by the two sides to subsume each other under one or the other enlightenment.

1. Introduction

The development of modernism toward political and scientific enlightenments resulted in two political directions. The first stresses democratic institutions guaranteeing freedoms of speech, press, belief, opinion, occupation, assembly, and equal rights. The second consists of liberation movements occurring in the third and the second worlds. The latter, called the communist block, claims to have realized liberation. Since both directions stem from modern Western consciousness, it is essential to decipher the basic ontological intentionalities which established the two main directions leading to contemporary political confrontations. They provide a format for the impetus of political enlightenment with democratic institutions, and scientific enlightenment with its call for

Lifeworld and Technology, eds. Timothy Casey and Lester Embree, Washington, D.C.: The Center for Advanced Research in Phenomenology and University Press of America, 1989.

liberation from institutions in favor of technical management of all socio-political affairs by competent experts. There are various affinities between these enlightenments, most apparent in the demand by scientists for "freedom" of research, and in the call by economists for "free enterprise" and complete submission to "market forces." These affinities will be treated as surface expressions, as second level phenomena, founded upon a more basic development of the modern ontological view concerning humans and nature. The ontological quest provides a focus capable of integrating approaches from various disciplines: sociological, juridical, political and scientific. The quest is not arbitrary. In his edited work, *Phaenomenologie und Marxismus*, B. Waldenfels credits thinkers such as K. H. Volkmann-Schluck, L. Landgrebe, and E. Fink for having pointed to an ontology that subtends the current socio-political predicaments.[1] Although without numerous references to their works, we shall follow their mode of digging to the roots.

2. Political Enlightenment

While analyzing modern consciousness, its interpretation of the world and its foundations, Eugen Fink reaches a conclusion that modern humanity has an awareness of itself and the world as if they were created out of nothing, as if there were no standard that could lend some stability to the world, allocate a place for humanity in it, thus releasing man from the bewitched circle of an incessant incrementation of power.[2] At first glance it would seem that Fink overstates his case. Yet a closer analysis of the modern ontological ground suggests that Fink was too mild in his assessment. As Tilo Schabert argues, modern consciousness is more radical and burst upon the West in a sudden revolution.[3] The revolution was directed against everything that was developed in the medieval period and against the Greek conception of a cosmic order.[4] One of the leading syndromes of this revolution, argues Horst Richter, is a humanly constituted "divine complex."[5] It is a consciousness that rejects both human inherence in nature and human dependence on transcendent powers. The result is a dramatic drive by the modern Western person to establish an omniscient and omnipotent position. This effort includes the constitution of an ontological consciousness.

The first step leading to political enlightenment appears in the devaluation of such conceptions as the human essence and the value of external nature. The rejection of an inborn human nature is one of the fundamental reasons for the "liberation" of the human from a belonging to a specific place in the cosmic scheme and indeed a proclamation of human freedom. Pico della Mirandola is the first to announce the new ontology. In his re-evaluation of the creation of human myth, Pico shifts the ground away from the classical and medieval views of human nature toward the conception of freedom and human self-creation. While elevating human dignity, Pico is rewriting the myth to make the human appear as a grand miracle (*magnum miraculum*). According to him, after the world was fully created and the standards and forms were exhausted, the divinity decided to create a "new son" (*novo filio*). Since the world was completely full, there was no specific place left for the human in the cosmic scheme, and since all standards were exhausted, the human was created neither in accordance with a "nature" nor for a specific place in the cosmos. The human is left without any form, without limits, and without any specific place. As a consequence, he must establish his own measures, standards, limits, talents, in accordance with his wishes and decisions. The human is neither divine nor demonic, neither mortal nor immortal; all this depends on his will. According to Pico's Testament, the human can become anything he likes.[6]

Not having any standards, any measure, the human is totally free to create all standards, to create his own nature and to be completely autonomous. While there are various modern expressions of human autonomy, extending through Kant and all the way to Sartre, Pico was the first to express this autonomy in its full ontological dimension. Human autonomy is not founded upon a will, a choice between two forces that would attract the human either from outside or inside. Every attraction would still tie the human to natural events and would assume that the human has inborn needs correlated to such events. To be attracted, in this sense, is to be enslaved by natural events and inborn needs. The modern ontology of autonomy has an opposite meaning: lacking any standards either from some creator or from nature, the human is autonomous and must create standards and rules and follow them despite external and internal causes and drives. Thus the human is conceived as a creator and giver of laws. To be in a position to establish laws without precedent either in myth or even nature and human form, is to be in a position to create them "as if out of nothing." Any standard to measure ourselves, to decipher who we are and can become, must first be created by us. The consciousness of the human ontological status as self-created is equal to a consciousness of the divinization of man. After all, only divine beings can be self-created.

The elevation of the human to a status of a sole source of laws is equivalent to the modern consciousness of human freedom, value, respect, rights and equalities. If this view is granted, political enlightenment must be essentially political liberation. Since every person is the source of standards and laws, there is an awareness that each person must have full rights to express his autonomy and to reject all traditions and their customs. Such an autonomy implies equality of all persons as law givers and a prohibition of domination of one person by another or a group without the person's consent. It must be noted that equality here has a different foundation from the one developed in classical Greece and the medieval period, where equality resulted from a conception of a common human nature, comprising an essence bounded by limits and a natural place. In modern ontology, all are equal not due to a common nature, but because each is an unconditional source of law. Thus no one can be more valuable, freer or higher than anyone else, and no one can impose his laws on others. Each, being a law giver, can only agree with others in the establishment of laws by mutual consent.

Given this context, institutions are created primarily not to guarantee that the individual adhere to the law, but above all to protect each individual's autonomy as a law giver, i.e., the right and duty to participate in the shaping of laws. Created freely by autonomous beings, the institutions secure, promote and maintain each individual's autonomy. Intent on strenghtening human autonomy, the thinkers of political enlightenment prevailed in establishing the division of governmental powers in order to preclude the government's rising above the institutions and thus establishing a rule of persons over other persons. Institutions should not become instruments of individuals or groups seeking their private interests and increase in power. As Martin Kriele argues, without the division of governmental powers democracy is not guaranteed. In contrast, such a division assures that laws can be changed democratically, that they are applied equally to all, including the appointed officials, and that the institutions are public and accessible to all.[7] Kriele also argues that democratic institutions cannot be founded either on private, natural interests or on the political power of a group.[8] His strict arguments against interests and power are convincing and show that democratic institutions are a result of the modern conception of human autonomy from which derive modern views of human value, respect, rights and equality.[9] In brief, autonomy cannot be negated without a contradiction.

Modern consciousness of ontological autonomy is found not only in political enlightenment but equally in other domains. The understanding that the human is a creator of standards, of laws and of himself, that he is a "divine beginning," stretches from the early Renaissance and moves through the romantic evaluation of man as an "original creator," a "divine genius,"[10] all the way to narcissistic self-glorification and Nietzsche's reach for the overman who, in a free exercise of power, creates and dooms even divinities.[11] Not even the so called "materialist" Marx rejects this ontology. According to him, by creating the material conditions of his existence, man also creates himself. The aim of history is to establish material conditions that will not only humanize nature and man, but also liberate the human from all material necessities for a free self-creation. This is a return to the original modern position that man is a source of standards and of himself. Perhaps the highest expression of the ontology of autonomy is found in Kant. Freedom is autonomy of a law giver.

3. Nature and Scientific Enlightenment

While political enlightenment elevates the human, scientific enlightenment devalues experienced nature. Prior to the modern revolution nature had beauty, goodness and value. If the human desired something, the desire was directed toward something only because the object of desire had an intrinsic good and value. But after the modern revolution, something has meaning, beauty, value and is good because humans desire it. I will it, not because it is good, but it is good because I will it. In principle, all that is meaningful, good, valuable, and beautiful, is on the side of the human. In this sense nature becomes "unmasked" and appears in its intrinsic meaninglessness, in its material reality. In his work, *La Nature*, Buffon pronounces the verdict: brute nature has no beauty. I, only I, make nature attractive and alive. Everything changes only through me, and a new nature flows from the hand of man.[12]

Although this elevation of the human and the degradation of experienced nature is founded on the ontological consciousness of human autonomy, such a consciousness is insufficient to guarantee human autonomy in nature. Man finds himself in a nature which far surpasses his powers and limits his autonomy. In order to achieve complete autonomy in principle, the human must become a law giver not only to himself, and in agreement with others to society, but also to nature. To conceive this all-encompassing autonomy, it is essential to recognize the divinization of humans in scientific enlightenment. Not to speak of Condillac, Rousseau, Descartes or Newton, this divinization already appears in the works of Galileo and Bacon. Galileo announces that nature is material and works in accordance with mathematical laws. Since humans are in a position to understand such laws completely, their comprehension of the material world cannot be distinguished from divine conception. In knowing the mathematical laws of matter, humans have wisdom equal to divinity.[13] If all material events are known from a divine vantage point, then human knowledge encompasses the material totality. According to Schabert, this means that human knowledge, respect and value are on par with divinity.[14] To speak with Galileo, man is so grand that he should be envious of himself.[15] While part of nature, humans place themselves above nature.

The modern elevation of human understanding of material laws to divine wisdom offers humans an instrument for changing material processes in accordance with human will. The very use of science not for pure knowledge but as a tool at the service of humans is one underlying intention that motivates scientific practice. For example, Bacon is not interested in regarding the material events for scientific understanding but primarily for the acquisition of power over nature. After all, Bacon's main work carries a subtitle *Man's Rule (De Regno Hominis).*[16] Human rule, an intentionality to dominate, requires a complete knowledge of material processes; otherwise man would not be in a position to subordinate reified nature under his will and standards. If there is something in the world that escapes human understanding and power, then humans are still in an inferior position in relation to nature. They are not yet full masters, not yet totally autonomous; they are subjected to nature. When all secrets are deciphered and dominated by humans, then humans will be fully autonomous, such that nature and the intrinsic human self will be reworked in accordance with the will and standards created by humans.[17]

Philosophers, including Descartes, were equally swept up by the will to dominate. Indeed, Descartes proposes the purpose of the new sciences to be tools for the subordination of nature under human domination. But when human knowledge becomes an instrument of domination, then it is already designated technologically.[18]

The subordination of nature under human scientific controls elevates the power of man. The incrementation of power is based on an instrumental conception of science, its technologization and the reification of nature. The reduction of the experienced world to matter allows everything to be an indifferent stuff that can be shaped and changed in accordance with human rules, wants and techniques. The terms "rules" and "techniques" refer to the human side of scientific enlightenment. The latter assumes that the only trustworthy method for science is mathematical; since the latter is not part of the materially conceived nature, it has to be a humanly constructed system of rules. Thus at the outset the rules by which matter is to be mastered are human creations. Consequently, the subordination of nature to mathematical rules is equivalent to domination of nature by human rules. The latter are theoretical, and at the same time, technical, allowing a direct movement of theory to technique.[19] The term "technique" has yet another meaning: concrete human activity in shaping the material environment. By combining the mathematical techniques with the concrete human activity, we obtain the structure that parallels the intentionality of domination. Every material state of affairs can be calculated mathematically; from the calculated material events humans can calculate and predict future material results flowing from, or being caused by, the events. The material events can be changed, can be reworked by human intervention in terms of human calculations, and from such reworked matter man can predict the "caused" results. The material events, calculated and shaped by humans, establish the conditions which yield expected results. The shaping requires not only a technological conception of rules and activities. The intention to dominate requires an additional step: not only can we calculate and construct material conditions and obtain predicted results, but we can reverse the process: it is possible to calculate material results, e.g., a novel process, such as superconductivity at a desired temperature, and construct material conditions, such as combining chemical mixtures to obtain the desired results. But the desired results do not exist in our environment. By producing an appropriate arrangement of material processes, we shall as a result produce the desired effect. Thus, the more results one projects, the more one must manipulate and dominate the material environment in order to achieve the results. The growing power to control nature is called progress. The latter assumes both the increasing projection of producible results and the growth of material calculations and techniques for the conditions that would produce the results. One begins to regard reality not as a "thing in itself" but as a possibility of being: possible conditions yield possible outcomes, and possible results require the calculation of possible conditions.[20] Thus one is no longer facing nature as it functions in its own right, but as it ought to function in terms of human will and rules.[21]

Within this context, modernity regards progress not as an acquisition of knowledge about nature, but as an incrementation of technical abilities methodological and physical–for the growth of power to master nature in terms of created projections. Indeed, as Hans Jonas argues, modern man is convinced that every technological application leads to new discoveries, and these discoveries lead to new applications, without end. Every transformation or reshaping of matter through technologies–which too are results of man's calculative activities–offer more possibilities and power for transforming more matter into projected results. Moreover, the shaping of matter into new techniques opens, in turn, a demand for more technologies and discoveries. If technical means make scientific discoveries possible, the latter lead to technological improvements and applications, *ad infinitum*. No achieved stage can be "perfect." Every stage reveals new and improved techniques and prospects. Thus, every achieved result becomes a means, i.e., the material condition, for the projection and fulfillment of new aims.[22] Here we can see the ontological domain structured by the intention of mastery. According to scientific enlightenment, this process elevates human power over nature and leads to a gradual liberation of humans from their natural surroundings and from their psycho-somatic afflictions. In brief, this process insures an increase of human autonomy, i.e., to be a law giver to whatever is still not under human mastery. As Spaemann suggests, modern ontology consists of mastery over and liberation from nature, man's realization of his ontological autonomy.[23]

4. Liberations

Contemporary political confrontations between the supporters of democratic institutions and the liberationists must be viewed within the context of the opposition that flows from modern political and scientific enlightenments. Holding strictly to the ontology of the latter, everything must be reduced to matter, capable of being transformed into material conditions for the achievement of projected results. With this technocratic logic, the human must be also reduced to material processes. Based on the ontology that man has no specific nature and hence can make of himself whatever he wants, one can draw a conclusion that at the material level the human is a product of various material conditions. Humans can be used as material conditions for establishing desired material results and it becomes possible to construct conditions capable of producing a "new man." Conjoined in one, we obtain a modern definition of man as *homo laborans* and as a "product." Science and technology can accept only one side of the definition: man is completely subjected to the influences of material forces. This view founds not only theories of behavioral modification, but also genetic engineering, the economic theories of Marxism-Leninism and capitalism.

Perhaps one of the best known and most pertinent examples relative to our topic is a sweeping manifestation of scientific enlightenment's consciousness among the scientific Marxists. The latter promise to establish "scientific socialism," founded on a consciousness that material conditions and technological power will be able to bring about a new man, and will be in a position to abolish all want and strife. Such an achievement requires a subjection of humanity to the reconstructive expertise of various technological fields and their "objectively" trained personnel. Of course, an interference in the work of the experts by the general public and its uneducated opinions is an unnecessary hindrance and should be excluded. In principle, no freedom of individuals is possible as long as humanity is functioning under forces that are not under man's control; freedom will appear only when man is in complete charge of all environmental factors. This means that it is only after the construction of the necessary material conditions that humans can become autonomous law givers in their own world. What distinguishes this "autonomy" from the autonomy of political enlightenment is the scientism of the former in its technocratic form.

To appreciate the results of such a scientism and its relationship to political enlightenment, we must articulate the meaning of liberalism. It is possible to distinguish three types of liberalism. The first belongs to scientific enlightenment and is based on the ontology of autonomy, with all the liberties, rights, values and equality flowing from it. These are secured by political institutions which, in principle, maintain pluralism and tolerance of opinions, political directions, moralities and divergent aims and the means for their attainment.[24] Within the context of this liberalism, a person lacks one right: abolition of democratic institutions that guarantee individual autonomy. The abolition would create a condition wherein man would not be subject to the institutions and the laws established by public consensus, but to the direct whim and power of others. Man would be dominated by man, and the result would be power confrontation.

The second type of liberalism, which dominates much of political and social thought, is *relativistic liberalism*. It has a tendency to account for human events by the use of the "scientific" model of conditions-conditioned. Thus human value, respect, equalities, rights and freedoms are expressions of various material forces. In this sense, the modern consciousness of autonomy, of man as self-creating, has no universal validity. At different times and in different places the socio-economic forces have produced other conceptions of who man is, and called for other political forms of rulership and institutions. Since the material forces change, human conceptions and institutions must equally change under the impact of such forces. Thus democratic institutions could vanish under different conditions. It would be impertinent to demand of other nations that they establish democratic institutions when neither the conditions are ripe nor democracy a necessary form of political organization. This claim, obviously, justifies any political system, and above all one that claims to be engaged in the "material reconstruction" of its productive-economic base as a means for a future liberation of its own population.

To what extent can such relativism be valid, i.e., can it be maintained as liberalism without contradictions? On the basis of relativistic liberalism, it is impossible to defend the necessity of institutions securing freedoms, rights, tolerance of others, since such factors are manifestations of temporal interests. This relativism precludes the possibility of judging socio-political and economic systems in which rights, freedoms and tolerance are abolished. After all, the material interests do not demand such institutions. Moreover, in case of changing conditions, the established democratic institutions would have to be irrevocably abolished. Relativistic liberalism would have to, and indeed does, accept this possibility. At the same time it contradicts itself by claiming that we must tolerate the political rights, views and freedoms of others. By allowing the abolition of democratic institutions, it allows the destruction of autonomy and all of its results. If every political orientation, including the one that abolishes democratic institutions, has equal validity, then the political orientation which abolishes such institutions has an equal right to do so. One must tolerate even despotic systems as expressions of material interests. All such systems would have to be regarded as "internal matters" and accepted as they are.[25]

Relativistic liberalism opens the door and is a correlate to scientific progressivism, and thus to *progressive liberalism*. Although the first intention of scientific enlightenment is tied into political enlightenment's efforts to achieve human autonomy, its principles are also at the root of the reduction of everything to material processes. As we saw, within its framework the human is a material condition and the conditioned, a means of production and a product. This is what leads to the subversion of the first intention to liberate the human from natural necessities; humans are left now with scientifically-technically structured material necessities. Instead of political liberation, the human reappears with a consciousness of material necessitation. Armed with this consciousness, progressive liberalism can regard the democratic institutions as products, as reflections of material and technical conditions.

Coupled with a dogma of material casuality, this liberalism opposes democratic institutions not only on the grounds that they are ideological expressions of material conditions and interests, but also by claiming that such institutions hinder scientific and technical planning for the establishment of necessary conditions for a scientific society and as means for final liberation from democratic institutions.[26] Following this logic, the progressive liberal can become a political technocrat and find himself above institutions. After all, he accepts a combined consciousness of (i) the technical and material processes causing all social life; (ii) the production of democratic institutions by a conjunction of material interests during a particular historical period; and (iii) the necessity of abolishing such institutions in order to transform society into a scientifically constructed and controlled process. The last point is crucial for progressive liberalism. If the political institutions remain, then the autonomous individuals in the population would have a right to decide, to challenge and to reject the activities of technocrats.

The political technocrats, armed with claims of scientific enlightenment, call for liberation from a democratic and sloppy management of public and private affairs, and their replacement by scientific and technical experts. What are the consequences of this liberation? In the main, humanity is not only currently immersed in ideologies, but is also ignorant of the "fact" that, like humans, ideologies are also products of material interests and conditions. Current humanity lacks the technical expertise to decide what is good for it and how to manage the complexities of socio-economic processes. Its voice is not only irrelevant, but, if permitted, it becomes a hindrance to progress. Such a human must be treated technically, as a material-productive means for the achievement of material progress. The political technocrat claims to know what sort of labor force is needed for progress as well as how the labor force is to be produced. As former President Reagan constantly insists, the labor force must be retooled to meet the needs of new productive conditions. The same language is used on both sides of the ideological fence, suggesting that the basic political concerns are not so much rights, freedoms, dignity and the public domain, but a fulfillment of human wants by submission to technically designed conditions and humans. In principle, then, the technocrats must be elites, unhindered by an ignorant public. Even if benevolently, on one side of the ideological fence, democratic institutions are seen as irrelevant, and on the other, they are replaced by political technocracy, and the new man must live in a society of nonconsensual man's rule over man.[27]

The progressive liberal is basically a political technocrat whose aim is to acquire power, to abolish democratic institutions and to appoint himself as a "leader of progress" with a vision of the true historical aim of a scientifically mastered nature and society. Accordingly, all liberation movements of this brand must be supported. Although such movements must necessarily abolish democratic institutions or, where they do not yet exist, prevent their emergence, they are nonetheless progressive since their aim is an unhindered establishment of material conditions that would yield, in some remote future, total autonomy. Yet at present, no deviating opinions, no rights, no political challenges are to be tolerated; they are outdated ideologies and only hinder progress. According to Landgrebe,[28] such a consciousness explains the shift from Hegel to Marx. In Hegel, only a divinity knows the direction, progress and ultimate aim of history. Yet the consciousness that regards the human as the master over nature (and given specifically the claim by some that they have expertise and know all the material processes) finds itself in a divine position. The "liberating" political technocracy, according to Lenin the political elite, assumes a divine role. This role allows the "liberators," the political technocrats, armed with "material science," to elevate themselves above ideologies and institutions, and employ any means, including humans, to direct history scientifically into the future known only to them. Thus mere mortals need not participate in nor can they challenge the decisions reached by political technocracies.

Since institutions guaranteeing freedom, tolerance, dignity and rights are rejected as outdated, disruptive of order and of the scientific-technocratic progress, the individual is subjected to rules established by self-appointed liberators and political technocrats. But neither the position, the views nor creations of the latter are subject to challenges. They are autonomous. We reach again the modern ontological consciousness: man is supreme master, a law-giver unto nature, except that now this law-giver wears a political-technocratic mask. This consciousness has by now become a world-wide phenomenon. It constantly insists that for the time being freedoms are "luxuries" one cannot afford; the priority of constructing the material future, including the need to produce an appropriate population, requires controls of all mass media for the sake of shaping the views of the public in accordance with the conceived needs of the political technocracy. Whenever the "scientific," progressive liberalism assumes power, wherever one finds liberation movements, one finds an immediate concern with dominating and, indeed, shaping consciousness and views.[29] The techniques of shaping mass opinion and the weight of such opinion, has become one of the pervading dangers of democracy in general.[30]

There appears a common denominator that cuts across opposing ideological postures: this is the support of "development," specifically in the form of technical "assistance" to the undeveloped regions of the world. The presumption is that the first priority is material, and that a progress in the material area will eventually lead to liberation and democracy. Meanwhile, the latter need not count, and the best form of political rulership is technocratic. This is most apparent in various third-world models, supported by both ideological camps. The major disagreement between such camps is geopolitical, i.e., which side will take control of managing the development and provide technical assistance. The decisive factor is not democracy and individual autonomy, but a technical control and use of populations for material progress. This legitimates either an abolition or a postponement of political democracy and the protection it offers the individual against arbitrary rule of man over man, and allows the political technocracy to rule in the name of "objective" material necessities. This position equally legitimates the use of humans as material means for developing the technical conditions necessary for progress. In this sense, the technocrats find themselves in a position above the public, with a right and power to treat humans as shapable material for the building of a "better" life. This mentality is not restricted to a geographic or political region. It is a pervasive structure both in the "democratically" structured societies and in societies already completely ruled by political technocracies. Current events in the Soviet Union testify to the power of an entrenched political technocracy.

Regardless of the promises by pure theoretical sciences to discover someday an all encompassing explanation, modern ontology precludes any final statement, since there is no practical limit to possible variations of materially combinable aggregations. Thus "scientific progress" cannot cease. Spaemann suggests that modern sciences, bent on technical domination of nature, cannot conceive of a completed science. If such a science were to arise, it would immediately abolish itself as science; it would cease to be necessary to the extent that it would be immediately surpassed by and incorporated into more progressive sciences.[31] There are good reasons for this conclusion. There are two distinguishable kinds of progress: (a) a process which reaches completion has a final aim. Building a house would be such a process. (b) An event in which every achieved end becomes one among complex means for other ends, and the latter becomes means for other ends. In the final analysis this progress has no other end apart from its own self-proliferation.[32]

Science, in its technological form, is the latter type of progress, and thus the political technocrats have no limitations to their activities. They are self-legitimating and can continue to rule under the guise of increasing demands for their expertise.[33] Moreover, nature in its reified status can be divided, constructed and reconstructed by diverse calculations, each providing a different access for shaping the environment, and each constituting a specific area of expertise. Hence, none of the experts, dominating one area, will accept limitations from others. There is no reason for any to cease their activities.[34] These factors imply one conclusion: increasing technologization of all areas of social life, which leads to an increasing domination by experts and thus to incessant subjection of the individual under an expanding complex of rules and regulations. It will be impossible to reach a state of total mastery of the environment and complete liberation. No discovery, after all, yields truth; every discovery is a catalyst for greater power and manipulation, for reproduction of nature along new models and constructs. In abolishing democratic institutions, the liberation movements cannot pretend to offer an autonomous life in the future. All signs point to an opposite result.

The opposite result is not an accident; it rather has its motivation in the very composition of modern consciousness. Indeed, the structure of the latter is its very motivation. Yet there is another, a much older, context. It has to do with the perceived political irrationality of "the people." Plato already argued against the wisdom of the populace in matters of import and complexity. If only the "philosophers" could lay their hands on the rudder of the state, if only they were in a position to direct the affairs of the peoples rationally, the human lot would be improved immensely. Of course, the condition layed down by the philosopher-kings is the "non-interference principle." People should not be permitted to make decisions in public matters; the "wise" know much better what is good for the population and thus their rule is most beneficial to all. To speak in modern jargon, the rule of science is the best rule. While following Plato in this respect, modernity adds one amendment: concrete techniques to structure the world and its populations "scientifically." Thus the rule by political technocracy is a modern variant of Platonic philosophical aristocracy. But in both cases the hindrance comes from democratic institutions; to rule the state scientifically, one must abolish such institutions. This is the condition under which the political technocracy would become omniscient ruler, backed by a vast material-technical power.[35]

5. The Crisis of Modernity

Apart from the crises of the modern era, analyzed by Husserl in his critique of Cartesianism and quantification of all events, the emptying of meaning through the cheap trick of objectivation, and by the existentialist critique of the emptiness, alienation and absurdity of the modern life, our explication reveals another crisis. Modern consciousness reveals a concrete crisis appearing in the difference between what could be called History I and History II.[36] Despite their differences, both histories appear and develop together. History I consists of the consciousness of humans as autonomous beings, capable of creating themselves, their political rules and, by reducing nature to matter, an environment in terms of humanly created standards, laws and will. History II consists of various efforts to realize History I. These efforts inhere in the technological progress of increasing human power to dominate nature. Since such domination has not yet been completely achieved–and, as we saw above, is in fact not an achievable aim it is postponed, delayed and constantly appears as something on a receding historical horizon. Moreover, the efforts to achieve the state of individual autonomy require not only more technology, but above all, increasing subjection to its demands for efficiency, speed, and to the expanding political technocracy. The more one strives to achieve History I, the more is one exposed to its frustration by History II. It is a crisis that is tensed between an ideal promise declared in constitutions and world organizations, and the impossibility of its fulfillment within the structure of modern scientific enlightenment.[37]

While the tension between History I and History II might be the driving force that subtends even the conception of progress, History II is undergoing a crisis of its own. There appears to be a growing mistrust of the technocratic abilities to solve human issues purely though technology, specifically when solutions in the material sphere have unforeseen, unsuspected, non-localizable and negative results.[38] In brief, the problems involve all facets of modernity. If technological processes, while increasing the power to dominate nature, create increasing problems proportionally to their own growth, then political technocracy must grow in power in order to solve the newly created problems. Thus the rule of political technocracy, its power, must increase. This is not to say that the increasing power of political technocracy can solve the problems; rather, the problems being created are overly complex and too big for any individual or a group. The result is an emerging cynical consciousness. It is becoming increasingly transparent that the technocratic power and rulership is growing both in extent and concentration. By promoting well being through rapid growth, political technocracy is also increasing its rule and power. Sloterdijk points out that this increasing power leads to a situation where even the political technocrats, the ardent "liberators," have developed a cynical consciousness. They know that they are in a feeble position to solve anything, and hence their rule rests on power alone. Thus their claims to fulfill the modern promise constitute a "false consciousness."[39]

6. **Postscript**

The development of scientific and political enlightenments opened the way for technological construction of the environment in terms of human designs and wants. Such a construction was deemed to be a movement liberating man from forces that were not under his control. Once the natural surroundings were transformed to serve human purposes, to function in accordance with humanly imposed designs, man would become a free law-giver to himself and to his environment. Such a movement combines both enlightenments, the political enlightenment emphasizing human autonomy and democratic institutions, and the scientific enlightenment stressing man's ability to master the environment. Yet the two enlightenments split apart and constitute two modern histories: the first supports democratic institutions and human autonomy, while the second demands that human autonomy cannot be achieved as long as the environment is not brought fully under human technical controls. The building of a true democracy and autonomy is a task best left to political technocracy and elite experts who must first establish the material conditions, enabling man to be free. This technocratic trend excludes public participation as a hindrance to the rule by experts.

The tension between the two histories is the driving force and the reason for various contemporary ideological clashes. Yet the latter are a surface manifestation of a more basic trend wherein the human effort to forge technical means of liberation led to a subjection of man by these very same means.

Endnotes

[1]B. Waldenfels, *Phaenomenologie und Marxismus* Bd.I (Frankfurt am Main: Suhrkamp, 1977), p. 9.

[2]E. Fink, *Traktat Ueber Die Gewalt des Menschen* (Frankfurt amMain: Vittorio Klostermann, 1974), p. 40ff.

[3]T. Schabert, *Gewalt Und Humanitaet* (Muenchen: Verlag Karl Alber, 1978), p. 50.

[4]Ibid., p. 94ff.

[5]H. E. Richter, *Der Gotteskomplex* (Hamburg: Rowohlt, 1982), p. 21.

[6]T. Schabert, *Gewalt Und Humanitaet*, pp. 168-170.

[7]M. Kriele, *Befreiung Und Politische Aufklaerung* (Freiburg: Verlag Herder, 1980), p. 42.

[8]Ibid., p. 172.

[9]Ibid., p. 76.

[10]T. Schabert, *Gewalt Und Humanitaet*, p. 244.

[11]H. Richter, *Der Gotteskomplex*, p. 55ff.

[12]T. Schabert, *Gewalt Und Humanitaet*, p. 207.

[13]Ibid., p. 140.

[14]Ibid., p. 141.

[15]Ibid., p. 142.

[16]Ibid., p. 93.

[17]Ibid., p. 195.

[18]Ibid., p. 201.

[19]K. H. Volkmann-Schluck, *Einfuehrung in Das Philosophische Denken* (Frankfurt am Main, 1965), p. 64ff.

[20]Ibid., p. 89.

[21]A. Mickunas, "Essence of the Technological World," in *Essays in Memory of Aron Gurwitsch*, L. Embree, ed. (Washington, D.C., University Press of America, 1984), p. 104ff.

[22]H. Jonas, "Philosophisches zur modernen Technologie," in *Fortschritt Ohne Mass*, R. Loew ed. (Muenchen: Piper Verlag), 1981, p. 81.

[23]R. Spaemann, "Unter welchen Umstaenden kann man noch von Fortschritt sprechen," In *Fortschritt Ohne Mass*, R. Loew, ed. (Muenchen: Piper Verlag, 1981), p. 106.

[24]R. Descartes, *Oeuvres Philosophiques*, Vol. II (Paris, 1967), p. 414.

[25]Ibid., p. 190.

[26]Ibid., p. 200.

[27]Ibid., p. 204ff.

[28]L. Landgrebe, *Phaenomenologie Und Geschichte* (Darmstadt: Wissenschaftliche Buchgeselischaft, 1968), pp. 132-3.

[29]K. H. Volkmann-Schluck, *Politische Philosophie* (Frankfurt am Main, Vittorio Klosterman, 1974), p. 174ff.

[30]Ibid., p. 180.

[31]R. Spaemann, "Unter welchen Umstaenden kann man noch von Fortschritt sprechen"? p. 110ff.

[32]Ibid., p. 97ff.

[33]Ibid., p. 107.

[34]N. Luhmann, "Gesellschaftsstruckturelle Bedingungen und Folgeprobleme des naturwissenschaftlichen Fortschritts," in *Fortschritt Ohne Mass*, R. Loew, ed. (Muenchen: Piper Verlag, 1981), p. 106.

[35]M. Kriele, *Befreiung Und Politische Aufklaerung*, p. 106.

[36]T. Schabert, *Gewalt Und Humanitaet*, p. 225.

[37]M. Kriele, *Befreiung Und Politische Aufklaerung*, p. 18.

[38]L. Gilkey, "The Religious Dilemmas of a Scientific Culture: The Interface of Technology, History, and Religion," in *Being Human in a Technological Age*, D. Borchert, and D. Stewart, eds. (Athens: Ohio University Press, 1979), p. 37ff.

[39]P. Sloterdijk, *Kritik Der Zynischen Vernunft*, Bd. I (Frankfurt am Main: Suhrkamp, 1983), p. 37ff.

FROM RECORDED SPEECH TO THE RECORD OF THOUGHT

Ivan Illich

Abstract: The alphabet is a technology that patterns writing and reading behavior. This patterned behavior can be viewed as a ritual. By analyzing a change in writing technique between 1140 and 1170, I will show a radical transformation of the ritual and of the meaning given it through the use of this technique.

. . . relying on hours that have lost their clock

"Je suis un peu lune et commis voyageur
j'ai la sp de trouver les heures
qui ont perdu leur montre. . . .

Il y a des heures qui se noient
il y a d'autres mangées par les cannibales
je connais un un oiseau qui les boit. . . ."[1]

Historians are free to assume very different stances in front of their object, which is the distance that separates us from the past. They can dig into the soil and follow the roots of the present. They can fancy themselves as camera-men engaged in a dive. They can explore a foreign country,

Lifeworld and Technology, eds. Timothy Casey and Lester Embree, Washington, D.C.: The Center for Advanced Research in Phenomenology and University Press of America, 1989.

notebook in hand. They can act as necromancers who evoke shadows and re-embody the dead. They can move away from their own, chosen epoch with their back to the present. They can also follow the example of the crab. Crawling backwards, the historian can keep an eye on the present, as it recedes, ever ready for rude surprises as he bumps into the pastness of the past. For such a one, the inquiry into the past turns into a revelation of the presentness of his own concepts and precepts. The past for him lies in a heteronomous *spime*.[2]

My commentary to the *Didascalicon* of Hugh of St. Victor has been undertaken with the intent to find leverage by which I could pry myself loose from certain assumptions about the "text."[3] I have lived through a transition from the page to the screen. I knew that it was upon me, when, for the first time, a student who had *listened* to me, turned to me and said: "Sir, I cannot *read* you!" Since then "the text" has become something like a verbal amoeba, a plastic term with no precise denotation and a multitude of connotations. The use of the term veils the shift from the book to the computer as the emblem by which we interpret self and society. It is this shift that I want to examine in the perspective of the shift from the monastic to the scholastic page around 1150.

1. The Alphabet as a Technology

An Augustinian Canon in Paris, Hugh was in his thirties and the Master of Saint Victor when he wrote the *Didascalicon* about 1128. The subtitle of the book is *de studio legendi*. For about four centuries, this book, next to Peter Lombard's *sententiae* and Gratian's *decretum*, remained part of every cleric's curriculum. The book has been often studied as a witness to the canon of readings done by the students of the *artes*. I want to look at it as a crucial witness to a turning point in the history of a technology that has shaped Western reality in a most profound way. The technology which I mean is, of course, the alphabet.[4] The 12th century inherited some twenty-odd letters, their basic sequence and some diacritical marks. It further inherited a set of tools that are used in writing: wax-tablets, parchment, the pen, the stylus and the reed. From the late-Imperial epoch, the middle ages inherited the book: the techniques of cutting of the scroll into sheets, of stitching these sheets together and of binding them between covers.

These technical elements were substantially left unchanged during the

12th century, but toward its midpoint they were integrated with a set of new techniques, conventions and materials. Some of these innovations were rediscoveries of skills known in antiquity: paper-making and cursive handwriting are two examples. Other techniques were invented anew: alphabetic arrangement of key-words, subject indexing, a kind of page-layout that is suited for silent scanning and the construction of the truly portable book. As a result, the inherited two dozen letters and their ABC sequence became part of a qualitatively new technology.[5]

I want to discover how these old and new technologies were integrated about 1130, and in which ways this integration led to a form of reading different from what was practiced a century earlier and a century later. I am reading the *Didascalicon* out of my interest in the symbolic effects that result from the use of a specific technology.

I want to understand what Hugh did when he read the book of his time: what habits and meanings were shaped by the interplay of the social skill of reading and the recording technique called "book" or *litterae* at that time.

I want to interpret what Hugh intended to do by reading: I want to understand what significance he gave to the use of alphabetic technology and reading habits within the context of the Canon Regular's life. I want to understand the symbolic effect that results from the age-specific habits within which an age-specific technology is put to use.[6]

2. From the Trace of Utterance to the Mirror of Concept

Before Hugh's generation, the book is a record of the author's speech or dictation. After Hugh, increasingly it becomes a repertory of the author's thought, a screen onto which he can project his still-unvoiced intentions.

As a young man, Hugh was introduced to an activity that we might call "monastic reading." He mainly listened to the book: he listened when he read it to himself, he listened when he chanted the responses in the choir, he listened when the book was read in the refectory or in the chapter room. And he wrote on the art of reading for people who would listen.

Those who in fact used the *Didascalicon* as their textbook were not listeners anymore. They engage in "scholastic " rather than in "monastic reading." And the technology they used was of a new and different order.

In Hugh's generation, the book is like a corridor, with the *incipit* as its main entrance. If anyone thumbs through the book and hopes to find a passage, he has as little chance to happen upon it as if he had opened the book at random. After Hugh, the book can be entered at random, with a good chance to find what one is looking for. It is still a manuscript, not a printed book, but it is already technically a substantially different object. The flow of narration had been sliced into paragraphs whose sum total now made up the new book.

What this means can be illustrated by an experience known to most of us today. Until into the late 1970's musical records could be re-played at will, but there was no easy way of random access to a specific beat or a melody. By 1988, however, not only groove-counters but also new electronic programming devices have become standard features on record players. In that same way, the book for the monastic reader was a discourse which you could follow, but into which you could not easily dip at the point of your choice. Only after Hugh did random access become a standard procedure.

Before Hugh of St. Victor, old books grow by mere accretion. During Hugh's life-time editing starts: legal decrees are ordered and collected; all known commentaries of Church Fathers to verse after verse of the Bible are assembled; Abelard gathers contrasting opinions on the same theological issue. Tradition is cannibalized and compiled according to the new editor's whim. Hugh is not one of them.

After Hugh's death pupils began to use these compilations. A new kind of reader came into existence who wanted to acquire in a few years of study a new kind of random acquaintance with more authors than a meditating monk could have perused in a lifetime. These new demands were both stimulated and met by new reference tools. The existence and use of these tools is profoundly new. Once these tools were invented they remained fundamentally unchanged, until the computer reached its 1980 stage. Now, within fractions of a seconds the computer command "search!" pinpoints a word in any file. A simple *macro* unlocks a diarrhea of citations or quotations without rhyme or reason. Authors think of themselves as operators or composers of texts. Only at this point does a new mutation of comparable depth set in.

These shifts from the recording of speech to the recording of thought, from the record of wisdom to the record of knowledge, from the authorities inherited out of the past towards the storage of promptly usable, well coined 'knowledge' can, of course, be understood as the reflection of a new mentality and economy during the 12th century. The changes in literate technique can of course be viewed as a response of the clerical trades to the demands of princes, lawyers and merchants. I want to look at this interaction between society and the page from the perspective of recording technology. How did the use of new technique foster new ways of conceiving reality?

In the hundred years after Hugh's death the estimated number of written accounts and legal charters in England multiplied by a factor of 50 to 100.[7] This alphabetical technologization of the word had enormous consequences on both a practical and symbolic level.[8] Described reality became legally more powerful than the witness's word: charters got the last word at court. In this interpretation of the *Didascalicon*, I want to explore how these changes have affected the axioms by which social reality emerges: how they affected the mind-set of generations.

3. From the Comment to a Story to the Story About a Subject

In Hugh's youth, learned books were either venerable "scriptures" (Bible, Fathers, Philosophers) or commentaries upon them. The teacher let the text of a "scripture" determine the sequence of his own exposition. *Ordo glossarum sequitur ordinem narrationis*. Sometimes the *glossa* was visually incorporated into the *narratio* upon which it was meant to comment. But glosses were, just as well, written on the margin or between the lines. This way of glossing is a visual consequence of the mental process of what is best called "monastic reading": an *auctoritas* says that "all forms of science are in the service of Holy Scripture." Nothing that went through the mind of the reader was deemed inappropriate as a commentary to the text. Texts grew out of tangents appended to older texts, which slowly were absorbed by them.[9]

During the first quarter of the 12th century a new kind of order appears on the manuscript page.[10] Inter-linear glossing becomes less frequent. Gloss and text with forethought enter into a new kind of marriage.[11] In the new order each is given its due: the gloss is subordinated to the dominating main-text. It is written in smaller letters. The way in which, henceforth, the two unequal partners are wedded betrays careful planning. The author himself becomes aware that the layout is part of a visual whole that determines the understanding of the reader.

Around 1260 Peter the Lombard[12] personally supervises the calligrapher who copies the dictation taken by his secretary. The book on which they work together is a verse by verse commentary on the Psalms. Before writing each page the calligrapher must have figured out how many words from the book of David fit on a page to make sure that enough space is left for the respective commentaries.

A new aesthetic sense expresses itself in this allocation. This new textual patterning of book pages had such a strong hold on the imagination that Guttenberg and his pupils did what they could to make its essentials survive into the age of print. This is not to say that earlier manuscripts are not, quite frequently, marvels of a harmonious spatial interplay of lines, glosses, miniatures and abstract lines with botanical creepers. But the new, abstract beauty obtained primarily by means of the layout of script is a result of a mid-12th century calculating approach to the use of letter sizes. It reflects a new pleasure of projecting mentally organized and quantified patterns of "knowledge" upon the empty space of the page.

In Lombard's commentaries key words are underlined with bright-red, mercury based colored lines. He does not leave it up to the reader to recognize quotes: he introduces primitive quotation marks to indicate where they start and where they end. In the margin, references to the source from which he quotes are given.

These ordering devices enable Lombard to make the text of Aristotle subservient to his own mental picture of its structure. But he is a man of his age. He does not dare to manipulate Holy Scripture in the same way. His commentary follows it line by line. In contrast, graphically the *sententiae* to Aristotle are no longer a collection of comments, tangents and excursions related to one verse after another. The commentary evolves visually as Peter's own line of thought is nourished by frequent references to a work of Aristotle. The gloss now has the purpose to bring out an *ordo*, which Peter has read into it. The learned book has ceased to be a sequence of commentaries that are strung like beads on the thread of somebody else's narration. The author now takes it upon himself to provide the *ordinatio*: he himself chooses a subject and puts his order into the sequence in which he will deal with its parts. The visible page is no more the record of speech but the visual representation of a thought-through argument.[13]

4. Ordinatio: Visible Patterns

At the same time a new written language comes into being that enhances the patterns created by this new graphic technique. By the early 13th century a terse sequence of glosses at the beginning of each chapter gathers the argument which will be treated. These arguments are put into a numerical sequence, *prima causa*, *secunda* . . . *quinta*. Standard rhetorical questions, like question marks, precede the conclusion of each argument. These questions are "marked" by a formula, which frequently is *obicitur* (which means "one might object"). An *auctoritas* or a quote or a "straw man" expresses doubt about the author's just-treated *argumentum*, and gives the author a chance to clarify his point of view with a *responsio*, his answer. These markers are highlighted on the page by special color. The reader immediately recognizes where the tempter or *adversarius* has been given the word. The visual marker shifts the task of perceiving the author's *ordinatio* from the inner ear to the eye, and from the rhythm of sound to a new factitious space. Reliance on this visual architecture of the *ordinatio* makes it increasingly necessary, when reading, to have the book under the eyes.[14]

5. Statim Inveniri: Instant Access

Beyond his concern with lay-out, the Lombard is also aware of a new time frame into which the act of reading has been moved.[15] He wants to lighten the burden of the student and speed up his reading. He wants to decrease the need for long leafing through the pages, and insists on chapter titles that allow the reader to find immediately what he is looking for. He arranges his *sententiae* in such a way that "for him who searches it be not necessary to turn the leaves of many volumes, but that he may encounter quickly without toil what he is after." Hugh insists on patience[16] and leisurely tasting[17] of what can be found on the page.[18] Peter wants to give his pupil all help that he can to locate with ease and with speed what he wants to read out of the book.[19]

The *scholastic* reading that Peter's generation invents stands in contrast with the monastic approach of a Bernard of Clairvaux, who insists that his monks engage in the hard labor that is necessary to "discover at great pains the hidden delights of Scripture" and warns them "to overcome the foreseeable difficulties that they will encounter in this search."[20]

6. Alphabetic

As already mentioned, the huge 12th century compilations are not yet reference tools in the modern sense. For example, the *Glossa Ordinaria*[22] is a huge collection of commentaries by the Church on the entire Latin Bible. Several monasteries collaborated over three to four generations.[23] Long sections surround the biblical text, and short comments are inter-linear. But there is no other cohesion in these many, partly overlapping compilations than the text of the biblical book. The *glossatores* just do not think of any more obvious way to edit, anthologize and order the material than by arranging the passages as footnotes to the Bible. Since all things that can be written about derive their meaning from their reference to an event told in the Bible, the most logical place to look up an option by Chrysostomos is not by subject but within this book. But techniques for searching the scriptures become common only toward the end of the 12th century.

Perhaps the most revolutionary of the new searching devices are those based on what we now call alphabetization.[24] We are by now so accustomed to encyclopedias and telephone directories that the use of the a-b-c-d-z sequence to create subject lists seems as "natural" to us as the use of the letters to spell a word. Scholars increasingly recognize that truly phonetic writing was a one-time invention made in Greece around 770 BC. The use of signs for both consonants (which are obstacles to breath) and for vowels (which indicate the color given to the column of air, which is spirited out of the lungs) constitutes a technique of immense social significance. It disembeds those societies that use it from the community of all other cultures. However, few scholars have realized thus far that the arrangement of names or of subjects in the order of these letters is a comparable technical breakthrough, which was made in the course of a generation. In analogy to the watershed which divides pre-alphabetic Greek oral culture from Greek culture under the aegis of letters and science, so it seems reasonable to speak of the pre- and the post-index Middle Ages.[25]

Without any application to indexing purposes the a-b-c sequence has shown an intense resistance to mutation. Since the dawn of history, it has remained, fundamentally, as fixed as the shapes of the Greco-Roman letters. The non-use of this sequence for subject listing is therefore a very remarkable and significant fact. It no more occurred to 85 generations of alphabet users to order things according to the a-b-c than it has occurred to the makers of the *Encyclopedia Britannica* to arrange articles by their references to the chapters and verses of the Bible.

The Greeks inherited the sequence from the Phoenicians. They revolutionized it by distinguishing consonants and vowels. Thereafter the Romans dropped some signs that were not needed in the Italic language, and added a couple, thereby expanding and contracting the a-b-g-d of the Greeks into the roman ABC. The letters were re-shaped in the 9th century.[26] In Hugh's childhood every schoolboy learned the alphabet, and knew it as well as the Our Father. But, on the whole, for 2,700 years, the sequence of the letters remained unchanged up to our time.[27] And yet, it was never used as an ordering device for the purpose of listing concepts or things.

Then, during the middle of the 12th century, came an avalanche of previously unthought of devices: indices, library inventories and concordances. They are all devices engineered to search and to find in books a passage or a subject that is already in the mind. The stuff from which these devices are made, namely the two dozen letters and their millennial order, remains unchanged. But the technical use to which the rote-memory order of the ABC is now being put is an essential ingredient in a conceptual revolution.[28] This revolutionary use of a trivial sequence rather than concrete events to order subject-categories is but one expression of the 12th century desire to create and to recognize a new kind of order. This desire has been well studied. It finds its aesthetic expression in architecture, law, economics and the new cities, but nowhere as clearly than on the page. The new page layout, chapter division, distinctions, the consistent numbering of chapter and verse; the new tables of content to the book as a whole; the new summaries at the beginning of the chapter referring to its sub-titles; and the new introductions, in which the author explains how he will build up his argument, are so many expressions of the new will to order.[29]

In each of them, a cultural impulse, a mental purpose and a graphic device combine to achieve something unprecedented. However, in no instance can we study the feedback of the technology on the mind-set as clearly as in the creation of alphabetic indices. The mental topology within which knowledge is henceforth pursued and which defines scientific procedures is henceforth heterogeneous with the topology within which Hugh's mind still moves. The author mutates out of the dictator of a story into the creator of a text.

7. Author Versus: Compiler, Commentator and Scribe

It is not the 12th century Latin corpus that is changed or enlarged. Hugh and his contemporaries still worked on the premise that all books that had survived from both Pagan and Christian antiquity were known to them. Those authors of antiquity who survived the dark ages in Arabic rather than in Latin were not yet available in translation to Hugh's contemporaries in Paris. The canonical authors were regarded by Hugh's pupils with the deference that he himself had shown towards them. But with the new ordering devices, they ceased to treat these old books as something that is simply reproduced or updated. Late 12th century writers digested them in a new way: no more as fodder for their own meditative rumination, but as building materials that could be used in the construction of new mental edifices.

Lombard and Gratian and the so-called *ordinatores glossae*[30] were still driven by the ideal of just re-ordering the Christian corpus. But only one generation later, after 1160, the creators of indexes were intent on drawing preconceived "contacts" from this corpus and making it accessible for the use of theological system builders, preachers or lawyers.

Hugh teaches so well ordered a way of reasoning that rather than "thumb through the pages" his pupil may use his recall as a finding device. The meditative reader discovers in the space of his own heart which thing or event refers by analogy to another. This is the reason why "We ought, in all that we learn, gather brief and dependable abstracts in the little chest of our memory." Only after Hugh's death do the sounding lines on the page fade and the page become a screen for the order willed by the mind. Rather than a means to revive a *narratio*, the theological and philosophical book becomes the exteriorization of a *cogitatio*, of a thought-structure. This *cogitatio* is not primarily the spoken memory of an event, but the thought-out outline of reasons.

The layout of the page in turn impresses this outline on visual memory. The page is broken up into paragraphs, each of them corresponding to a *distinctio*, a distinct point of view. Markers call attention to the sequence of *distinctiones*. While titles had been occasionally used in Antiquity, after Isidore they became uncommon and in the 13th century they return with a vengeance. In the *Summa* of the 13th century, the references to the author's intent, such as *quaestio, obicitur, respondeo dicendum* hold the exposition itself in their grip. Dictation becomes almost impossible for the theologian unless he refers to his notes.

With more clarity than anyone before him, Bonaventure distinguishes the tasks in the intellectual authorship of a book.[31] "There are four ways of making a book. There are some who write down alien words, without adding or changing a thing, . . . called scribe, *scriptor*. There are those who however . . . add something, however not their own additions . . . called compiler, *compilator*. Then, there are those . . . who annex their own for clarification, who are called *commentator*. But he who writes both what comes from himself and from others, but uses the alien mainly as annex and confirmation, he is the one who ought to be called author, *auctor*."[32]

8. Lay Out

The ear, so far, had distinguished the voice of the dead author from the voice of the reader. Now, the visual articulation of the page called for a new distinction between kinds of persons who each contribute a special feature to the texture of the page. The new layout reflected not only the will to use visual articulation as a means of interpretation; it also led to the first attempts at textual criticism. Three hundreds years before printing made it possible to "establish" a critical edition, attempts to untangle the threads from which the visible text was woven became common. That which comes from the author who writes his own, is distinguished from its shaping and ordering by other clerks. Inexorably "the" text of a book comes to be distinguished from the version in this or that other manuscript. This distinction became important, long before the technology of printing would make it possible to establish the text in a critical edition.

9. Illuminatio Versus Illustration

The meaningful layout which addresses the eye has to compete for attention with the illustration on the same page.[33] Five functions that have been attributed to the ornamentation of christian manuscripts can be distinguished.

Cassiodorus (ca 485-580) considered illustrations like solemn vestments, which by their beauty provided the word incarnate on the page with an appearance worthy of its dignity. He knows the book as a sacred object, which is carried around with great solemnity during the liturgy, is honored with incense, illustrated by a special candle, and is kissed on the painted initials before and after the loud reading of the passage these pictures mark off.[34] The book is an object of worship: just as gold and gems are used to embellish its covers, its pages are covered with colors.

Besides honoring the word with clothes that befit it, the illustration had a didactic purpose.[35] It was meant to show to the bodily eye of the simple-minded what exceeded the grasp of his mind. While "listening" to the written word, the picture nourished the imagination.[36] A technique used for this purpose was the so-called Exultet-Rolls,[37] commonly used on the northern slope of the central Alps, and named after the *exultet*, a solemn long chant sung by the deacon during the Easter night. Its haunting, unforgettable melody is still close even today to old synagogue patterns. The deacon stands behind a lectern while the Easter candle is blessed, intones bible verses with their commentaries from greek Church Fathers, and retells the story of Salvation from the flight out of Egypt to the resurrection of Christ. And as he sings he unwinds a roll on which text and pictures follow each other. When you look at the roll, the pictures are upside down, and their subject precedes the story by many lines. The reason for this curious feature is a simple one: the deacon is meant to unroll the scroll of parchment over the lectern, so that, while he tells the story, the people who listen, in front of him, can contemplate the illustrations.

But illustrations are not only meant to help in the instructions of the simple-minded.[38] As a third function they are meant as exegetical and heuristic cues to prompt the monastic reader. They are intended as non-verbal vehicles for the same revelation which letters transmit as sounds.

Hoc visible imaginatum figurat illud invisible verum, stands in the caption of one of these miniatures: "this visible effigy represents that invisible truth."[39] Another caption is even more explicit an admonition "here thou shalt contemplate, that . . . ," *hic erat contemplandum*. An Irish code of the 12th century suggests that method that connects word and image: "What this picture allows you to grasp with the bodily senses is that which you should bring forth spiritually."[40]

Fourthly, the early Medieval miniature is conceived as an accompaniment that supports the sound given off by lines when the reader moves through them. The miniature is meant to bring out the sparkle of the page's voices. It does not have in any way the purpose of the graphs or charts in a modern textbook, where such devices reduce the subject-matter to an abstract clarity for which language is too clumsy. Medieval illuminations invite the mumbler to fall silent in adoring what no word could express. Nor are the pictures, like photographs, meant to document a fact, or to provide evidence for the matter that is discussed in the text. Miniature and lines interlace ear and eye in the perception of the same symphony (*ridon le carte*).[41]

Finally, the illustration of books before the 13th century has a practical now often forgotten–mnemonic purpose. Hugh speaks of reading as a journey. He advances physically from page to page. The ornaments that line the rows of letters place the words into the landscape through which this journey leads.[42] On no two lines does the reader meet up with the same view, no two pages look alike, no two initial "A-s" are identically colored. The foliage and grotesques in combination with the lines re-enforce the power of remembrance: they support the reader's recall of the *voces paginarum* in analogy to the scenery of the road that brings back the conversation which took place on the stroll.

Modern reading, especially of the academic and professional type, is an activity performed by commuters or tourists–no more that of pedestrians and pilgrims. The speed of the car and the dullness of the road and the recognition of billboards put the driver into a state of sensory deprivation, which continues when he hurries through manuals and journals once he arrives at his desk. As the tourist is equipped with a camera, so the student today reaches for the Xerox to keep a snapshot souvenir. He is in a world of photography, *illustrations* and graphs, which put the memory of illuminated letter-landscapes beyond his reach.

During the 12th century the intrinsic coherence of the line and its illumination dissolves.[43] As the line is made into a building element of paragraphs, the miniature turns into a circus of phantasy-creatures, often a jungle which threatens to invade and to overpower the alphabetic component of the page.

Bernard of Clairvaux in several sermons attempts to exorcise this sensual intrusion of spirits gone wild. He recognizes that illuminations have their place in the prayer books provided by bishops in the pastoral care of lay-people.[44] He insists that flamboyant ornaments addressing the senses can foster a sensual devotion or piety in people whose spirit has not been awakened. But he strongly objects to the illustration of manuscripts meant for the spiritual readings of monks, whose entry into the cloister has led them beyond the this-worldly crowd. In 1134 a general chapter of Cistersians demands that even letters be written plainly and in one color only.[45] This veto went unobserved; it was too starkly in contradiction with the trends of the times.

10. The Portable Book

The miniature of the late Gothic manuscript frequently is an independent work of art. Sometimes the letters seem to be the mere frame for the pictures. In the age of growing urban wealth the book becomes an object of private ownership: its miniature paintings enhance the status of the owner's wealth.[46] This privatisation of the "book" as a physical entity could not have happened without a further set of technological breakthroughs.

The Bible, as one single bulky object, was still unknown in the 12th century. It still was, as it always had been, a collection of tomes. And more often than not, the tomes were of different size, written for different usage, assembled, occasionally, as the canonical collection of sacred books. "The Gospel" lay on the ambon to the right hand of the celebrant for reading by the deacon, and the Epistle lay to the left to be read by the lector, and these were separate tomes. The Psalter was open on a lectern in the middle of the choir. "The Pentateuch" was usually under separate covers from "the Prophets."

Two merely technical reasons supported this functional division of the Bible into differently stapled and differently ornamented objects. The pages available were too bulky and heavy to carry a complete Bible, and the letters used were too large to fit the text into it. Only in the 13th century were handwritten letters shrunk to a size that made it possible to fit all of the Bible into one volume. The intensive use of word-abbreviation contributed to this compression of script. But, even so, a 13th century Bible rarely weighs less than 10 pounds. Other techniques had to be created to make the step from the luggable to the truly portable Bible.

Crucial in this advance was the preparation of a new, thinner "virgin" parchment: it was made from the skin of unborn sheep, carefully tanned by using alum and delicately smoothed with fine grained pumice stone. The technique was costly, fit for many jewels. Barely had this new technology become known, when paper re-appeared into Europe. Through the Egyptians, antiquity had known and used the papyrus-reed based material,[47] but it disappeared for almost a thousand years. During this time Arabic traders who ventured into Transoxania acquired the art of papermaking from Chinese artisans. And this new competitor for parchment reached Paris–via Spain–just in time for the opening of its University. Scholastic studies could hardly have come into vogue without this new, cheap, light writing surface.

The reduction of letter size, the reduction of page-weight and new abbreviations were, by themselves, insufficient to make the book portable.[48] A new way had to be found to stitch small paper sheets in such a way that they open fully in the hand of the reader. Also, a new, flexible cover had to be engineered for a book which, for the first time, was made to be held, rather than opened on a support.[49]

By 1240, in its essentials, the book was much more like the object we take for granted than it was like the book into which Hugh gazed.

Endnotes

[1]Passage from a poem by the Chilian poet Vicente Huidobro (1893-1943) who wrote mostly in French. "I am part moon, and part traveling salesman/my specialty is that of finding those hours that have lost their clock . . . believe me, there are hours that have drowned, and others eaten by cannibals; I even know of a bird that stills its thirst on hours. . . ."

[2]"Spime" was used colloquially by Einstein in lieu of space-time-continuum.

[3]The present essay, written at the invitation of Professor Lester Embree, grew out of the second of the seven draft-chapters for a book which deals with the activity and the meaning of "*lectio*" in the mid-decades of the 12th century.

[4]For an introduction to the study of the alphabet as a technology, and its history as such, see Walter Ong, *Orality and Literacy: The Technologization of the Word* (London: Methuen, 1982), and also the annotated bibliography in Ivan Illich, and Barry Sanders, *ABC: The Alphabetization of the Popular Mind* (San Francisco: North Point Press, 1988).

[5] M. Rouche, *Des origines de la Renaissance* (Paris: Nouvelle Libraire de France, 1933), (Vol. 1 of L.H. Parias: *Histoire generale de l'enseignement et de l'education en France*). The first four carefully edited and illustrated volumes that deal with successive stages of written culture in France. Both the techniques of writing and reading and their relevance to reading are examined by the contributors.

[6]Ludolf Kuchenbuch, *Schriftlichkeitsgeschichte als methodischer Zugang: das Pruemer Urban 893-1983* (Hagen, Fernuniversitaet, Lehrbrief 2 1982).

[7]M.T. Clanchy, *From Memory to Written Record: England 1066-1307* (Cambridge, MA: Harvard University Press, 1979), reviews what is known

about the increasing use of written records, and the manner in which this increase both reflects and enhances the perception of the relationship between individuals and society. Clanchy stresses the effect of literate documents on everyday life and, on purpose, does not focus on parallel developments in literature, science, and philosophy.

See also: A. Murray, *Reason and Society in the Middle Ages* (Oxford, Claredon Press 1978), a social history of literate competence and of the powers it confers on the literate from 1050 to about 1300.

[8]I. Illich, *Schule ins Museum: Paridros und die Folgen* (Bad Heilbronn, Klinkhardt 1984), pp. 53ff.

[9]B. Smalley, *The Study of the Bible in the Middle Ages* (New York, Philosophical Library 1952). Also see B. Smalley, *The Gospel in the Schools c. 1100-1280* (London, Hambledon). It was in the school of Laon that, for the first time, the whole Bible, rather than just Psalter and Pauline letters, was expanded. "Peter Comestor is the earliest recorded lecturer on the Gospels on Paris. He ordered that the traditional glosses in relation to the next text and the various senses of the Scripture distinguish between the senses with unprecedented precision: he was arguably the only commentator of the period–and for a long time to come–to suggest a theory of historical development."

[10]N.M. Haering, "Commentary and Hermeneutics," in *Zum Stand der Forschung ueber die Entwicklung der Glosse und des Kommentars in den mittleren Jahrzenhnten des 12. Jhs.* (Benton/Constable, 1982), pp. 173-200.

[11]R.H. Rouse, and M.A. Rouse, "The Verbal Concordance of the Scriptures" (*Archivum Fratrum Praedicatorum* 44, 1974), pp. 5-30. The concordance to the Bible "did not evolve over generations, but was invented and perfected through careful tinkering and adjustment in less than 50 years" (p. 5).

[12]J. Chatillon, "Les ecoles de Chartres et de St. Victor." *La scuola nell' Occidente Latine nell' alto Medio Evo,* Settimana di Studio 19, Spoleto 1972, vol. 2, pp. 795-839. At the recommendation of St. Bernard, Peter had come to St. Victor in 1139 to follow Hugh's commentaries on Scripture.

[13]W. Goetz, "Die Enzyklopedien des 13. jahrhunderts" in *Aeitschrift fuer die deutsche Rechtsgeschichte 2,6* 1936, pp. 227-250. Encyclopedias of the new kind appear first around 1225 (Bartholomaeus Anglicus), followed in 1240 by *De rerum natura* by Thomas Chantimpre and Vincent de Beauvais, who digests 2,000 books by 450 authors.

[14]Malcom B. Parkes, "The Influence of the Concepts of *Ordinatio* and *Compilatio* on the Development of the Book" in J.J.G. Alexander, and M.T. Gibson, ed., *Medieval Learning and Literature: Essays Presented to Richard William Hunt* (Oxford 1976), pp. 115-141. "Thirteenth-century scholars paid close attention to the development of good working tools based on scientific principles. The drive to make inherited material available in a condensed or more convenient form led them to recognize the desirability of imposing a new *ordinatio* on the material for this purpose. In the thirteenth century this led to the development of the notion of *compilatio* both as a form of writing and as a means of making material easily accessible. Compilation was not new (it is implicit in the work of Gratian and Peter Lombard); what was new was the amount of thought and industry that was put into it, and the refinement that this thought and industry produced. The transmission of these refinements on to the page led to greater sophistication in the presentation of the text" (p. 127).

Vincent of Beauvais elevated *compilatio* into a literary from: "In working out his scheme, with commendable humility he followed the example of the Almighty . . . *ut iuxta ordinem sacrae scriptura, primo de creatore, postea de creaturis, postes quoque de lapsu et reparatione hominis, deinde vero de rebus gestis iuxta seriem temporum suorum, et tandem etiam de iis que in fine temporum futura sunt, ordinate disserem.* In the *Speculum naturale* he follows the chronological order of the six days of creation given in the Book of Genesis. At the lower end of *ordinatio* his procedure was influenced by the *modus definitivus* of his own age. Since, according to Alexander of Hales, . . . *apprehensio veritatis secundum humanam rationem explicatur per divisiones, definitiones, et ratiocinationes,* Vincent achieves the subordination of his material by dissecting his *auctoritates* and redeploying the diverse materials into discrete, self-contained chapters. In the *Speculum naturale* the third, fourth, and fifth day of creation give him the opportunity to review all that was then thought about minerals, vegetables, and animals. By dividing his work into books and chapters he is able to include as many as 171 chapters on herbs, 134 chapters on seeds and grains, 161 chapters on birds, and 46 chapters on fishes. In the *Speculum historiale* by the same process of redeployment into discrete units he includes such material as the account of the ancient gods, and the biographies of leading authors' of antiquity accompanied by extracts from their works–all subordinated within the framework of universal history. In all, the *Speculum maius* is divided into 80 books, and 9,885 chapters: it is the classic example of the principle of *compilatio* which emerged in the thirteenth century, divide and subordinate'" (p. 128).

[15]Peter Lombard compiled his Sentences: *"ut non sit necesse quaerenti, librorum nume ositatem evolvere, cui brevitas, quod queritur, offert sine labore"* (PL 192, 522). Ibid: *"ut auten, quod quaeritur, facilius occurat, titulos quibus singulorum librorum capitula distinguuntur, praeisimus."*

[16]DB V,2 p. 96: *Sic et mel in favo gratius, et quidquid maiori exercitio quaeritur, maiori etiam desiderio invenitur.* "Thus also is honey more pleasing because enclosed in the comb, and whatever is sought with greater effort is also found with greater desire" (DT, p. 121).

DB IV, 1 p. 70: *Contra, divina eloquia aptissime favo comparantur, quae et propter simplicitatem sermonis arida apparent, et intus dulcedine plena sunt.* "The Sacred Scriptures, on the other hand, are most fittingly likened to a honeycomb, for while in the simplicity of their language they seem dry, within they are filled with sweetness" (DT, p. 102).

[17]The word "*sapientia*" for the Latin speaker evokes *sapere*, "to taste." The search for wisdom connotes the aspiration towards something of ultimate, exceptional taste: *hoc ergo omnes artes agunt, hoc intendunt ut divina similitudo in nobis reparetur, quae nobis forma est, Deo natura, cui quanto magis conformamur, tanto magis sapimus* (DB II, 1, p.23). However, note, that the expression "in good taste," to which we are accustomed, appears in Spain only during the last decade of the 15th century, and has probably been coined by Queen Isabel.

[18]DB III, 13 p. 62-63: *Considera potius quid vires tuae ferre valeant.*

aptissime incedit, qui incedit ordinate. quidam, dum magnum saltum facere volunt, praecipitium incidunt. noli ergo nimis festinare. hoc modo citius ad sapientiam pertinges. "Consider fellows who wishing to make a great leap of progress sprawl headlong. Do not hurry too much, therefore; in this way you will come more quickly to wisdom" (DT, p. 96).

[19]R.A. Rouse, and M.A. Rouse, "Statim invenire: Schools, Preachers, and New Attitudes to the Page," in R.L. Benson, and G. Constable, eds., *Renaissance and Renewal in the Twelfth Century* (Cambridge, MA: Harvard University Press, 1928), pp. 201-225.

[20]Parkes.

[21]R.H. Rouse, *"L'evolution des attitudes envers l'autorite ecrite: le developpement des instruments de travail au XIIIe siecle"* in *Culture et travail intellectuel dans l'Occident medieval*, p. 115-144 (Revue des Sciences Philosophiques et Theologiques 66, 1982).

[22]The expression "*ordinaria*" is a designation used only since the 14th century.

[23]See B. Smalley, *Recherches de Theologie ancienne et Medievale 9 (1937) 365-400.*

[24]*Horst Kunze, Ueber das Registermachen*. Leipzig (1966, 2 ed.). M. A. Rouse, and R. Rouse, "History of Alphabetization," in *Dictionary of the Middle Ages,* vol. 1 1982, pp. 204-207. Homer G. Pfander, "The Medieval Friars and Some Alphabetical Reference-Books for Sermons," in *Mae 3* 1934, pp. 19-29. This is on fifteen reference books of 13th through 15th century friars that provide alphabetical lists of biblical allusions, patristic quotations, stories, analogies and references to other sources for literally thousands of sermon topics.

[25]Obviously the reference from an index to the text could not be made by "page-number" until printing–250 years later–insured that in every copy of a given edition the first word on the first line would be, e.g. "Rome." Reference had to be made by "book, chapter and verse."

H.J. Koppitz, "Buch," in *Lexikon des Mittelalters*, vol. 2, p. 802ff. Sets of leaves had been occasionally numbered in antiquity. In order to insure correct sequencing in the bindery, they were commonly numbered in the right lower corner of a new set and called *custodes*, watchmen in the Middle Ages. The so called *reclammates* were written on the following first page of the set.

[26]David Ganz, "The Preconditions of Caroline Minuscle," *Viator* 18, 1987, pp. 23-43. The production of normative script . . . required a *grammar of legibility* which facilitated the understanding of these texts by clerics whose native speech was germanic and whose latinity was seldom Augustian.

[27]This statement is not entirely true if we were to include in it the *devanagari* of India and other systems of syllabic notation that have been derived from it.

[28]Albertus Magnus, *De Animalibus* in Opera Omnia, Auguste Borgnet, ed., (Paris, 1890-99), vol. 12.433, excuses himself for listing animals in alphabetical order because this is not worthy of a philosopher, "*hunc modum non proprium philosopho esse,*" cit. Rouse, 1982, p. 211.

[29]Like all of the new techniques which converged in the new 13th century layout, introducing summaries also have a history. The so-called *accessus*, in which a commentator prefaces a text with tidbits about the author's life, the content of the work and its utility to the reader, goes back to commentators on Greek philosophers. See Edwin A. Quain, "The Medieval Accessus ad Auctores," in *Traditio* 3, 1945, pp. 215-264. The new summaries, however, describe the outline of the book or chapter which they introduce.

[30]B. Smalley, "La Glosa Ordinaria," in *Recheres de theologie ancienne et medievale* 9, 1937, pp. 365-400.

[31]M.-D. Chenu, "*Auctor, and Actor, Autor,*" in *Bulletin du Cange III*, 1927, pp. 81-86. Sallust and Suetonius occasionally designate the person who dictates a book both as *auctor* (from *augere* in the sense of bringing

forth something like a statue, building or writ) and *actor*, like the agent of anything, the maker. During the next 1000 years, the one Author, the Author of Holy Writ, namely God, gave to the word a transcendent meaning. Simultaneously *auctoritas*, the dignity which a statement carries because of its intrinsic worthiness, reflected some of its meaning back to auctor-actor. By the end of the 13th century, author stresses the *authenticity* (p. 83) of the orginator.

[32]Here, slowly, the modern concept of the author as the authentic originator takes shape. See A.J. Minnis, *Medieval Theory of Authorship: Scholastic Literary Attitudes in the Later Middle Ages* (London: Scholar Press, 1984).

[33]Christle Meier, and Uwe Ruberg, ed., *Text und Bild: Aspekte des Zusammenwirkens zweier Kuenste in Mittelalter und frueher Neuzeit* (Wiesband: Reichert, 1980).

[34]J. J. G. Alexander, *The Decorated Letter* (New York: Braziller, 1978).

[35]L. Gougaud, "Muta praedicatio," in *Revue Benedictine (de Maredsous)* 42, 1930, pp. 168-171. Where he quotes, among others, the following witnesses to a tradition of manual catechists, having a vocation to preach by hand:

Cassiodorus (AD 575): The whole body begins to preach: what a happy invention, *felix inventio*: to preach to people with one's hands, to open their tongues with one's fingers, silently to mediate salvation to mortals, and to struggle against the subreptious attacks of the devil with the writing cane and with ink: *manu hominibus praedicare, digitis linguas aperire, salutem mortalibus tacitam dare et contra diaboli subreptiones illicitas calamo atramentoque pugnare.*

[36]Beda Venerabilis (AD 735): We make images for those who cannot read so that *posset opera Domini et Salvoris nostri per ipsarum contuitum discere*, so that they may learn about the works of our Lord and Saviour by the contemplation (of its picture).

The Synode of Arras (AD 1025) orders the use of pictures so that those who cannot see the Lord in the scriptures may contemplate him in the outline of pictures: *illiterati, quod per scripturas non possunt interueri, hoc per quaedam picturae lineamenta contemplantur.* The stress lies on the two distinct ways of being illuminated by "seeing" in the letters and "contemplating" in the painting.

[37]H. Douteil, [Easter Preconium (MS Biblioteca Lat 9820)] *Exultet Rolle.* Bibl. Vat. Graz, Akad. Druck und Verłagsanstalt (1975). Guigues Cartusian (AD 1136), *si ore non possumus dei verbum manibus praedicamus* (cit. Gougaud). *Honorius Augustodienensis (AS 1130), ob tres autem causas fit pictura: primo quia est laicorum litteratura* . . . the first of the three reasons for which a manuscript is illuminated is the provision of a "literature for the layperson, that is the illiterate."

[38]S. Lewis, "Sacred Calligraphy. The Chi-Rho Page in the Book of Kells," in *Traditio* 36, 1980, pp. 139-159. "The page . . . intended to confront the eye with an awesome array of Christological and Eucharistic allusions consonant with the liturgical usage of the book" (p.51). "An enormous erudition underlies each illumination."

[39]Christel Meier, "Zum Verhaeltnis von Text und Illustration bei Hildegard von Bingen," in A. Brueck, ed., *Hildegard von Bingen, 1179-1979. Festschrift zum 800. Todestag.* (Mainz, 1979), pp. 159-169.

[40]For references see M. Smeyers, *La Miniature (Typologie des sources du moyen age occidental,* Fasc. 8), (Turnhuot, 1974), pp. 96-101.

[41]Dante, *Divina Comedia, Purgatorio* 9, 82.

[42]H. Jantzen, "Das Wort als Bild in Der fruehmittelalterlichen Buchmalerei," in *Ueber den gotischen Kirchenraum und andere Aufsaetze* (Berlin, 1951), pp. 53-60.

[43]In the 12th century, the picture addresses the onlooker no more by speaking to him, but by being another kind of literature. See Thomas Aquinas, IV sentent. 1,III,9 2,2,3. *Fuit autem ratio institutionis imaginis in Ecclesia: primo ad instructionem rudium qui eis quasi quidam libris edocentur:* For the illiterate in the mid 13th century, the picture is already

conceived of "in the likeness of a book."

[44]*Scimus namque quod illi* [the Bishops] . . . *carnalis populi devotionem, quia spiritualibus non possunt, corporalibus excitant ornamentis. Nos vero, qui iam de populo exivimus* . . . (Bernard of Clairvaux, *Apologia ad Guillelmum abbatem 12, 28; Sancti Bernardi opera*, ed. J. Leclereq 3, pp. 104-5).

[45]*Litterae unius coloris fiant et non depictae. Status Capitulorum Generalium Ordinis Cistercinesis ab anno 1116 ad annum 1786*, J. M. Canivez, ed., vol 1, (Loewen, 1933), stat. LXXX, p. 31.

[46]Paradoxically, new wealth and a new spirit of voluntary poverty converged in their demand for portables. C.H. Talbot, "The Universities and the Medieval Library," in F. Wormland, and C.E. Wright, eds., *The English Library before 1700* (London: The Athlone Press, 1958), pp. 76-79, suggests that the mendicant friars were another major force behind the reduction of booksize. They wanted to carry them on their constant travel.

[47]Theodor Birt, *Das antike Buchwesen in seinem Verhaeltnis zur Literatur, mit Beitraegen zur Textgeschichte des Theokrit, Catull, Properz und anderer Autoren* (Berlin, W. Herz 1882). Reprint: Aalen, *Scientia* (1959), pp. 223-285–deals with the papermaking in Antiquity.

[48]Graham Pollard, "Describing Medieval Bookbinding," in J.S.G. Alexander, and M.T. Gibson, eds., *Medieval Learning and Literature, Essays Presented to R.W. Hunt* (Oxford: Claredon, 1976), pp. 50-65.

[49]Both feats were accomplished. Few engineered objects have gone through a change as radical and rapid. See Pollard.

TEXTS AND THINGS: HOLDING ON TO REALITY

Albert Borgmann

Abstract: Texts can be divided into traditional (literary, musical, and architectural) and technological ones (information on tapes, discs, and chips). Traditional texts require an engagement with community and reality to be realized, i.e., to attain their proper place in the world while technological texts are self-realizing and so disengage us from community and reality.

What you have got from your forebears,
Acquire it if you are to possess it.
Goethe, *Faust* I, 682-83.

The purpose of this essay is to articulate common intuitions about the new era of technology that we are moving into. The entry to this new epoch is marked by expressions such as the Second Industrial Revolution, the Micro-electronic Revolution, the Postindustrial Society, or Postmodernism. My concern is to bring a measure of clarity and consensus to our understanding of the present epochal changes. One may hope to succeed in this if one's inquiry is guided by both circumspection and insight.

I will begin with circumspection and go on to suggest that it can deepen into insight. Broadly speaking my enterprise fits into the scholarship that places our crisis in the history of the media that have distinguished human

Lifeworld and Technology eds. Timothy Casey and Lester Embree, Washington, D.C.: The Center for Advanced Research in Phenomenology and University Press of America, 1989.

culture and sees it as the last stage in the sequence of orality, literacy (chirography and typography), and electronics.[1] More particularly, my endeavor is an attempt at elaboration and correction. I am concerned to bring out real and communal virtues of literacy that are commonly overlooked by scholars and mortally threatened by the electronic media.

To serve circumspection, I will center my discussion on the notion of the text. That discussion, I believe, can provide a modicum of agreement and a point of departure for deeper and more contentious issues. The latter arise when we take up the question of how texts are related to things or to reality. My emphasis will here be on the antecedent and preparatory issues. But I will suggest how I think the story should be continued and concluded. Of course the notion of a text, its interpretation and status in reality, is itself very much in need of circumspect clarification. I want to provide it in a ready and intuitive way. As I proceed, I will, drawing on diverse scholarship, make this sketch more precise.

Let me explain what I call a text by beginning with a narrow and well-established concept of a text and then by widening it to include more and more items. Horace's odes constitute a text. So do Galileo's treatises or the Constitution of the United States. Any coherent composition or record that is available in written or printed language is what I would like to call a *literary text.*

Further I want to include musical compositions or scores under the heading of text and even architectural plans or blueprints. Literary, musical, and architectural plans, typically on paper, I will generally call *traditional texts.*[2] These have close counterparts in recordings contained on tapes, discs, and microelectronic chips. Such electronic records can contain, of course, literary, musical, and architectural texts. But often they contain richer information such as musical or dramatic performances, computer programs, scientific or technical information. Let us call these latter sorts of recordings *technological texts.*

Next let me informally draw a distinction between texts and things or events. Texts in one respect are of course things as well, things of paper or plastic, books, tapes, discs, etc. But a score compared with a performance and a blueprint compared with a building are extremely thin and desiccated things.[3] A text is to a thing or event as a seed is to the plant or freeze dried food is to the meal. A text needs to be animated, unfolded, or expanded to take its proper place and have its real effect in the world. The process and result of awakening and actualizing a text I will generally call *realization*. We have a number of specific and customary words to designate realization, viz., reading, reciting, playing, performing, building, and making in the case of traditional texts. Some of these we use for the realization of technological texts too; in addition we use terms like projecting, showing, running, and others.

I can now state the task of the essay more precisely. It is to articulate the novelty of the dawning postindustrial era by working out the different ways in which traditional and technological texts are realized. I want to suggest that a culture where technological texts are prevalent sponsors a style of world appropriation that is generally inferior to one where traditional literacy, the skill and the practice of reading and acting on traditional texts, is central. I do not urge a return to traditional literacy. We must rather regain and relocate it so that it stands in a secure and fruitful relation to the technological texts.

I begin with the traditional text, more particularly with the conjunction of permanence and contingency that characterizes a traditional text. To see its permanence consider Horace's ode 30 of the third book of odes that begins

Exegi monumentum aere perennius . . .

Horace wrote it sometime in the twenties B.C. on papyrus or a potsherd.[4] But the original record has not been preserved. In fact the oldest surviving manuscript is from the ninth century. Thus there is a break of nearly a millennium in the physical or material continuity of the Horatian text between its composition and the first extant record of it.[5] And yet the copy of the ode that I have before me, Wickham's Oxford edition of 1877, contains the very text of the ode. In spite of the vagaries of traditions and editions, the present identity and integrity of the ode is undiminished and unchallenged.

Horace must have sensed the apparent immortality of his poetry, for the first line of the ode we have been considering says in translation: "I have erected a monument more lasting than bronze . . . Clearly the perfect durability of Horace's art is due to its textual character. Sculptural works of art, as Horace implies, are more likely to be diminished or destroyed. One of the most famous in classical antiquity was the statue of Athena that

Phidias had fashioned for the Parthenon on the Acropolis in Athens in the late forties of the fifth century B.C. As with Horace's manuscript, the original is lost, but a dozen or so copies of it have survived.[6] Yet the latter have not preserved the undiminished identity of Phidias's Athena.[7] It has been irretrievably lost.

Texts in a certain sense are independent of the physical medium in which they are embodied.[8] A Horation ode is one and the same whether it is scratched on a shard, written on papyrus, or printed on paper. What makes such medium neutrality possible? Literary texts have a formal identity and permanence due to their digital or notational character. Let me explain this formal aspect of texts informally. A literary text must be written in distinct marks or inscriptions, each belonging to a certain letter or character. The number of different characters is limited, and in the case of each particular mark it must be unambiguously clear what character it belongs to. Thus in an alphabetic script there may be two dozen letters, and for a particular mark it is clear what letter it belongs to, e.g., whether this particular mark is an "e" or not.

The first requirement, then, for a text to be digital or notational is roughly that it consist of distinct components, e.g., of letters. A second requirement is that there be rules of composition or arrangement.[9] These may in part be very simple. They may say that components that are horizontally continuous form a compound, e.g., words, and that the compounds are arranged in a sequence of some sort, e.g., in lines running from left to right and listed from top to bottom.[10]

If the distinctness of the components and the rules of composition are understood and respected, copies can be made ad infinitum without damage to the identity of a text. Of course particular letters may show imperfections; an "o" may not be entirely closed, a "t" may not be fully crossed. But as long as copyists recognize the marks as an "o" or a "t," they will in their copies restore what negligence has clouded or time has worn away.

Alphabetic writing is not the only digital or notational system. Among the many kinds let us consider two more, musical notation and architectural drafting. The former notational system is more complex than alphabetic writing since it consists of very different kinds of characters, the staff, clefs, key signatures, notes, and so on. Musical notation is, none the less, perfectly digital. Architectural drafting is less rigorously formalized.

The components, i.e., walls, doors, windows, etc., are not perfectly distinct, and the rules of composition or arrangement are not fully formalized. Still it approximates a notational scheme.

Traditional texts, then, appear to have an immutable identity. But we must be more precise. Texts as things or physical objects are clearly perishable. In fact many millions of texts printed on acidic paper are crumbling every day.[11] Texts as bearers of meaning or as semantic objects are not immutable either. Their meaning may change or be lost altogether. It is as a purely syntactic or formal object, based on a notational or digital scheme, that a text appears to be immutable and "more lasting than bronze."

Texts in this formal sense can be thought of as fixing the identity of things or events, of plays, concerts, or buildings. The power of texts to secure and stabilize things is more informally and broadly visible when one considers the rise of literacy out of orality. Walter Ong has emphasized the liveliness and precariousness of oral thought and expression in a primary oral culture. Stories, theories, doctrines, and expertise are alive fully and only in the practices of a community. They are shaped by the limits of memory and kept alive through exercise and repetition. They change their shape constantly as the reality changes that they respond to.[12] Writing and especially printing convert these living practices into permanent, self-contained objects that are closed off and distanced from reality. Important parts of life are captured and fixed in texts. And what finally guarantees the rigor and permanence of a text is neither its physical embodiment nor its semantic content but its pure form.

The contrast thrown into relief by Ong is real and illuminating; but as regards contemporary culture, it is misleading too so long as the problem of realization and the distinction between traditional and technological texts remain in the background.[13] Still, Ong helps us to put the project before us in still another and suggestive perspective. Just as the transition from orality to literacy led to a restructuring of social life and an impoverishment of certain communal practices, so the transition from traditional texts (or literacy) to technological texts (or electronic information processing) is about to transform our common order once more and to threaten radically our remaining communities and engagements with reality.

Yet even this way of putting things is skewed; for, as I now want to show, literacy does not merely carry a residual orality and thereby communal practices with it; rather the realization of traditional texts enforces specific practices. In fact, even the pure and apparently permanent and immaterial form of texts is informed by *contingency*. In speaking of the contingency of the text I use the term in a negative and a positive sense. Negatively, I mean the opposite of the immutable, formal, and explicit, i.e., the changeable, informal, and implicit. Positively speaking, I use contingency to denote the mode of existence or presence that is distinctive of traditional texts. It is roughly a conjunction of authority and communal practice.

The contingency of texts is most obvious from the fact that no text speaks by or for itself. The texts of Linear B would not have been more easily decipherable had one of the writers of this ancient form of Greek left us a treatise on deciphering, written in Linear B.[14] An illiterate person is not helped when given a book on how to read. The legibility of texts obviously requires a living and continuous community that sustains and hands down the skill of reading.[15] What are the characteristic dimensions of the practices of teaching and reading? Reading is so deeply rooted in the distinctively human capacities that an investigation of reading runs the risk of getting lost in the general and difficult problems of perception, cognition, action, and language. Reading is not a strictly novel or magic human skill but the adaptation of more general abilities to a specific task. It is therefore worth emphasizing that to concentrate on the characteristic dimensions of reading, I will presuppose the normal skills of cognition and speaking. These presuppositions, of course, mark a fluid boundary between reading and other human capacities and practices.

To begin let me set aside the broader social and cultural aspects of reading instruction and habits and consider more narrowly the question of how a reader, confronted with marks on paper, comes to understand a text. It seems at first that the basic reading skill is a very limited one and requires only that one master the conventions of notation, i.e., the rules that match printed symbols with sounds, the so-called grapheme to phoneme correspondence rules. On the basis of that skill a reader, it would seem, could convert a text into a vocal or subvocal representation, and once a text is vocally present, the normal command of language could take over.

Though this is a misleading account of today's normal reading, it provides a helpful starting point for the examination of the contingency of texts. It allows us to see that fluent reading is an eminently bodily and communal skill. Reading is the laborious integration of steps and subskills into a fluid and powerful mastery that engages us intimately with the world about us.

The rise of fluidity out of discontinuity begins unnoticed at the lowest physiological level of perception. What we practice and experience as the smooth gliding of our glance along the printed lines is, microscopically considered, a series of overlapping jumps, so-called saccades, and fixations on words, the former requiring 20 to 50 milliseconds, the latter 200 to 250 on the average.[16] Saccades and fixations are native to all seeing and need not to be learned. They must be adapted to reading, but practice is the only teacher for that task.[17]

It is otherwise at the level of converting letters into sounds and integrating sounds into words. Teaching has a time-honored place in the acquisition of this skill. Acquisition is particularly difficult in English since its orthography incorporates much etymology. Hence the relation of letters to sound is complex. The first task is to group letters into graphemes of one or more letters so that each grapheme corresponds to one phoneme. The how of such parsing is perfectly achieved by a fluent reader, but hardly understood by the theorist. In "shepherd" p and h belong to different graphemes, in "periphery" they belong to one and the same. And of course one and the same grapheme can correspond to very different sounds. Consider "gh" in "rough" and "ghost." Setting aside rare or unpredictable cases, there are 100 to 200 grapheme to phoneme correspondence rules.[18] Finally, the reader must integrate the sounds into words and words into sentences that exhibit the usual patterns of intonation.

Normal reading does not end in this way. To be sure most readers can read aloud material that they fail to comprehend due to absent-mindedness or lack of expertise. But even if reading went no further, it would exhibit a most intricate skill and testify to strong communal practices. The skill is eminently embodied. It resides in the body or it does not exist at all. It cannot be replaced by following the guidance of manuals because it is the basis of any reading of instructions. And reading is a deeply embodied skill. The labor of learning a demanding skill runs up sharply against the bodily obstacles of slowness, weakness, and awkwardness. Accordingly, the command of a demanding skill gives us a distinctively human power and grace. It greatly extends our world appropriation.

The skill of reading also has a uniquely communal dimension. The community preserves and conveys in itself the conventions of how letters agree with sounds. And every student of reading is initiated by a community into this agreement. The possibility of texts rests on this twofold agreement. The members of a community agree that letters are to agree with sounds in a complex and principled way. Moreover, the literate community provides for every new member the help and discipline required to enter into this agreement.

Normal reading typically issues in the comprehension of a text. The scholarly inquiry of reading up to the level of word recognition is voluminous and detailed. Where it proceeds from word recognition to text comprehension it chiefly reveals perplexity and ignorance. Some of the fluent reader's normal achievements are quite evident. But how they are accomplished and hang together and how much of the process of reading and comprehending they leave uncharted remains in the dark.

To comprehend a text is, at any rate, to absorb its meaning. But since comprehending is rarely the same as memorizing, the reader must be able to convert the meaning of a text into a form other than the exact wording the text had. Finally, accomplished readers seamlessly integrate the meaning of a text with everything they know. Upon learning from a newspaper article that President Nixon had resigned, competent readers absorbed this information in such a way that it transformed and fit with what they knew of Nixon's biography, of the Watergate affair, of Constitutional history, of political theory, and much more.

What this sketch of comprehension suggests is that reading is appropriating. It is the explication and animation of the structure and meaning of the text, the integration of that structure and meaning with one's understanding of the world, and thereby the enrichment of one's world appropriation.[19]

It is impossible to describe in a general and illuminating way all the kinds of reading, comprehending, and integrating of texts into one's world.[20] The varieties of texts and of the circumstances of reading are too manifold. But it is possible to discover certain normative dimensions that order all sorts of reading. In fact the consideration of reading from word recognition or mere vocalizing to deep comprehension traces just one such dimension.[21] I now want to retrace this dimension to refocus the inquiry on the realization and contingency of texts.

In doing so I will principally urge that no sharp and abiding separation between a text and its realization can be made. It is this unity of text and realization that I mean by the contingency of the text. Texts are always realized some way. What we normally call the text in contradistinction to its readings, interpretations, performances, etc. is simply the currently thinnest, shallowest, or most general realization.

As we have seen, in considering reading, at the bottom of all texts is a notational agreement, conventions that specify the constituent elements and the rules of composition of a particular notation. But only in artificial languages such as symbolic logic is the notational agreement made explicit and explicitly. The elements or primitive symbols are listed and the rules for compounding them into well-formed formulas are specified.[22] It is all the more interesting that even such rigorously formalized languages are inescapably enmeshed in the contingent practices of the wider world. The intrication of formal languages in the world is evident in their being both excessive and incomplete.

They are excessive in that any one formal language exhibits more than its pure form. A formal language by its arrangements draws in particular and contingent ways on our worldly experiences. (I will consider the conventionality of the elementary symbols when I turn to written English.) Consider, e.g., the following formulas of symbolic logic.

(1) E x K f x g x

and

(2) (E x) (f x · g x)

They symbolize in Polish and Peano notation that there is at least one x that is both f and g. What they have in common can be diagrammed by representing classes as labeled circles, classes possibly with common members as intersecting circles, and the existence of at least one member of a class by placing an x in the appropriate circle. We can then represent what (1) and (2) have in common by way of a so called Venn diagram.

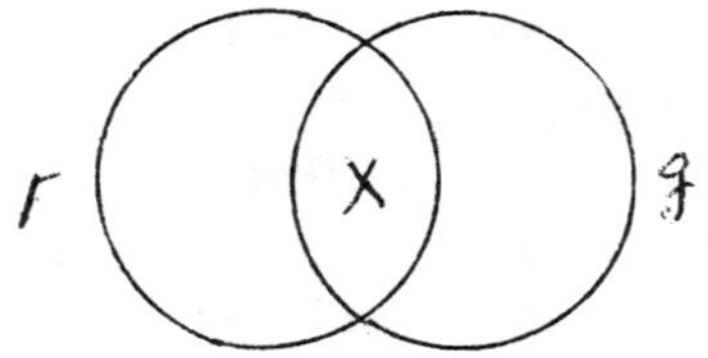

It is apparent that a Venn diagram presents the form of (1) and (2) no more purely than the discursive statements themselves. Each representation draws on certain things and practices and holds others at bay and so serves some purposes well and others less so. The Polish notation represents conjunction as a processor that takes two values and produces a third. The Peano notation presents it as a fastener or glue that distinctively makes two things into one. The Venn diagram represents it in a way suggesting enclosure in two corrals. Venn diagrams are intuitively obvious, but silent or implicit about inferences and arguments. For discursive notations the opposite is true.[23]

At the same time, formal languages are incomplete in that they presuppose and anticipate practices that cannot be formally stated. This is obvious and well-known from the fact that a formal language requires an informal meta-language in which to be defined and made intelligible. Moreover, every formal language has indefinitely many particular applications and instantiations which, it goes without saying, cannot all be committed to paper. But it is precisely the command of these particular extensions that is the mark of whether someone has mastered the formal language or not. Such mastery is an open-ended, textually unspecifiable practice though we can in practice determine through tests whether the mastery has been attained or not.[24]

The formally and textually statable part of the notational agreement that underlies written English is very small. The stability and intelligibility of English texts rests largely on the continuing practices of literate communities. The weight and intricacy of this practice is evident from the length and difficulty of the apprenticeship one has to go through to attain the full franchise in this community. It normally takes fifteen to twenty years to become a competent reader of demanding texts.

The implicit and historically changing agreements and practices that delimit and sustain traditional texts are what I mean by the contingency of texts. Though one can hardly overestimate the complexity and richness of this contingency, it is possible and helpful to explicate one major dimension of it, the dimension that is fixed by the polarity between thin and thick or shallow and deep realizations of a text. It is evident that the continuum between a thin and thick realization of a text is analogous to the continuum between the mere vocalizing and the deep comprehension of a text.

We can take our point of departure from the evident fact that texts on paper have some sort of privileged position in the contingency of texts. This privilege can be understood when texts are seen as memoranda and legacies. Historically, texts began as reminders or memoranda, as commonly accepted signs that this is so and that its being so should endure in our memory. A record of so many bushels of wheat delivered is a sign that helps us to remember that event.[25] But it is a help only. A record is not self-enforcing. It rests on the mutual agreement and trust that this mark shall forever stand for that event.[26]

Given trust and agreement, a sign does not have to overpower or indemnify; it does not have to be materially powerful or valuable but can be reduced to the merest physical trace as long as it serves well in reminding everyone who has entered into the agreement through schooling and affirmation. Hence the physically thin and attenuated character of texts. Signs serve well if they are readable and parsimonious at once.

To be recognizable and memorable the elements have to be distinctive. They should stand out as artifacts and not be confused with decorations or naturally occurring patterns and traces.[27] Hence the artificial shape of letters. For a text to be parsimonious, the gigantic number of their larger, i.e., compound, signs should be reducible to a small stock of components and a few systematic rules of composition.[28] This norm, as we have seen, is directly met only by formal languages and indirectly by strictly phonetic alphabets. Written English approximates it only in part. Although considerations like these allow us to situate and illuminate the nature of texts, they cannot reduce it to the convergence of nontextual components. Writing and texts are original cultural events. As with language, we can bring certain of their aspects into relief but only at the cost of letting others recede into darkness.

The aspect that mainly concerns us here is the contingency of texts. Texts begin as physically attenuated and commonly agreed upon memoranda of past events. They are delicately balanced between the objective and the subjective, between the enduring and the passing. In reminding us, texts command us and constitute something more powerful and precise than our memories. In being reminders only, texts depend on our affirmation and animation for their life and efficacy.

Of course writing, when fully developed, does more than remind. Through their relative permanence and precision, texts on paper allow us to conceive, elaborate, and record structures more artful and complex than those possible in a primary oral culture.[29] Thus the author of a traditional text may entrust us with a thing or event that had no existence prior to the text. But what the author gives us is not the thing or event itself but an instruction to realize the thing or event. Here the contingency of the objective and subjective takes the form of one between personal creativity and communal practice. Authors draw and depend on the things and practices of their community for the intelligibility of their texts. But they also see or hear something without precedent and signify and entrust it to their community.

Legacies and instructions are in principle incomplete and depend on trust and tacit agreements to be realized. The contingency of legacy and realization is delicate and shifting. A shift can be recognized as we look from one contingency to another. But it is impossible to fix the movement of these shifts absolutely by dividing the contingencies into their contingents. The pure form of the text is the chimera that tempts us to fix the text and set it apart from all realizations.[30] To be sure, there is always a presently thinnest realization of a text, viz., the text on paper. It constitutes the minimal agreement that the members of a community have bound themselves to, and indeed it is the agreed upon bond that can unite communities across millennia into a community of communities.

But the text on paper is neither the pure text nor the real text. Like the text of the formal language of logic or mathematics, written English too is both excessive or impure and incomplete, i.e., less than fully real. It is not pure because, as a physical object, it bespeaks through its spelling, font, and face a certain time and place of origin. Moreover, to come to our attention, a text must not only be realized on paper but also and minimally through silent reading. Such reading suffuses the text with particular communal practices and individual experiences and expectations.[31]

The text on paper is not the real text. The real text is the fully or deeply realized text. The text on paper possesses its privileged generality at the cost of being thin or shallow. All those who agree on nothing more than that these marks on paper constitute the score of Bach's *B Minor Mass* agree on relatively little. To the extent that they confirm nothing more than this agreement they must each abstract from all the depth and power that the *Mass* in various ways has for them. Moreover, it is impossible to say absolutely how much the little is that all who recognize a text on paper do agree on. Relatively speaking we can say that those who recognize the text of Maclean's "A River Runs Through It," all living in the late twentieth century, agree on more than those who, tenselessly put, agree on the text of Horace's Ode III, 30 since they are spread over millennia and very different cultures.[32]

Texts differ in depth of realization not only according to their provenance but also according to their kind. The realization of Bach's *B Minor Mass* reaches more deeply into the world than 2 + 3 = 5; the latter in turn is realized much more widely than the former. And any one text can be realized more or less deeply or widely at any one time. In the case of Bach's *Mass*, common agreement may go no further, if that far, than the details of the score. There are great differences of realization when it comes to the number of performers, the soloists, the choral and orchestral sound, the instrumentation, the articulation, and the tempo.[33] And finally, I fear, there is very little consensus today on when and where the *Mass* is properly to be performed and on how it is to inform our personal and political life.

This picture of texts as invariably realized at various depths may seem relativist since it appears to deny all moorings. But texts do in fact have their center of gravity, their point of concentration and orientation. They have them in what we may call the momentous realization where legacy and realization enter into one moving and festive event. The eloquence of the text and the vigor of our practices of realization are necessary conditions for such an event.[34] In the modern era, texts have tended to fall silent and our practices have decayed. The general response to this decline has been the attempt to divide and conquer the contingency of text and realization. The rise of technological texts has been the most significant result.

Technological texts are divided into presentational and instrumental ones. The former kind has a telic priority. Both kinds, however, have self-realization as their characteristic trait. To explain this trait, let me begin with presentational texts. In a general sense, they sprang from the same root as did traditional texts, from the desire to surpass the limits of human memory and to secure a sort of permanence to things or events that would naturally be ephemeral. This is how we may think of the daguerreotypes and Edison's sound recordings. More generally, technological texts are designed to overcome the frailty of human nature and to accomplish powerfully and assuredly what humans make and do in limited and fallible ways.

From the start, technological texts were not directly writable and readable by human beings. They required more than paper and pencil to be recorded, and the record or text needed a reproducing machinery to be properly visible or audible to humans. The reason is obvious. Technological texts were intended to be fuller and richer records than traditional texts. It takes a machine to gather the corresponding amount of information in real time and to reproduce it accordingly.

It seems at first that presentational texts are paradoxically related to reality. On the one side, such a text is more perfect the more fully it replicates the reality to be recorded. But on the other side and at the limit, a perfectly faithful replica is no longer a text or record of something but the thing once more, i.e., a duplicate. Thus the perfect text is no text at all.

The usual resolution of the paradox is reduction. Thus a recording of the *B Minor Mass* does not reproduce the room where it was recorded, the chairs, the performers sitting on them, and the instruments they held. It reduces the performance to mere sound and records and allows us to reproduce the sound only. Even a film is a reduction of reality. It abstracts from the smell and the temperature of the set. The visual reality is reduced to one angle. In viewing a film we cannot move right or up to obtain a different view of the scene.

The ideal resolution of the paradox, however, is not reduction but *pure appearance*. Pure appearance represents an ideal of completeness that overcomes the limits of reduction without leading us back to reality. We can roughly follow the path from reduction to pure appearance by considering the history of recorded music. The first recordings were both reduced and impure. They failed to capture the full range of dynamics and frequencies and the spatial unfolding of sound, and they did contain unwanted noise. The development led by way of high fidelity vinyl records, stereophonic sound, and magnetic tapes to compact discs, as the bearers of the pure and full sound of music. We may think of videos as examples where the visual dimension is added to the acoustic albeit in a flat and grainy way. A fine-grained visual quality is technologically feasible today, and for tomorrow, futurists expect three-dimensional and tangible projections.

There is another sense in which pure appearance is the telos of the realization of presentational texts. What typically appears before us when such texts are played or projected today is not the record of some real thing or event that could have been witnessed directly by one who was at the scene of recording. Sound recordings are touched up, refined, and synthesized. What is reproduced from the record could have never been heard at this quality and as one piece during the recording session. Similarly what an observer sees on a movie set is disjointed, composed of false facades, and surrounded by incongruous settings.

Still, in these cases there is a residual tie between the record and some recorded reality. The appearance that is truly pure in the sense of being independent of all debts to original reality is achieved when a presentational text is based on a notation and a grammar. A notational text is a digital text. But a technological text can be either technically or grammatically digital. A compact disc, containing the recording of a musical performance, is technically digital. It records an essentially analog object, the continuous waveform to which sound can be reduced. The waveform is decomposed by the recording machinery into discrete magnitudes, 40,000 per second, which in turn are expressed in binary code and stored as a series of bumps on the disc. The player reads the code, reconverts it into an analogue and feeds it into the speakers via the amplifier.[35]

A digital recording is noiseless and permanent. Thus it possesses the permanence and medium independence of a notational text. But as a formal object, a series of O's and 1's, say, it is grammatically or originally unstructured. It is just the copy of a waveform. Nowhere in its digital history is there a stage where the structure of the music was formally explicit, i.e., the key, measure, sequence of notes, kinds of instruments, etc. However all this is in principle recoverable in cases where a computer is used as an instrument and music is synthesized from the ground up.[36]

Some computer scientists and futurists believe that there is a grammar not only of the sounds of music but also of the visual world, at least of the significant aspects and happenings in it, including human action. There is in fact some understanding of the grammar of the laws of spatial arrangement and perspective in architectural drafting, of the movement of cartoon characters and of airplanes over terrestrial terrain, and of the ways in which simple solids tumble.[37]

But all these phenomena are evidently artificial and restricted. Is there a prospect of extending and refining the grammar of visual and auditory shapes to encompass the behavior and appearance of human beings? As often, technological imagination runs ahead of technological reality. Max Headroom, we are told, is the software impersonation of reporter Edison Carter. But the software is in turn impersonated by actor Matt Frewer.[38] Bill Gates, chairman of the Board of Microsoft, however, believes that if in twenty years "you want a certain movie star to be sitting on a beach, kind of being lazy, believe me, you'll be able to do that . . . . If you like Bill Cosby, then there will be a digital description of Cosby, his mannerisms and appearance, and you will build your own show from that."[39]

Pure appearance is not the immediate telos of instrumental technological texts but perfect self-realization is. By *instrumental texts* I mean very nearly all technological texts that are left when we have subtracted those that directly present us with a pure appearance of an auditory, visual, and eventually perhaps tactile and olfactory sort. But to focus our discussion, let us consider an example of an instrumental text, viz., the kind of program that guides numerically controlled machine tools. Machine tools are lathes, milling machines, cutting tools, and the like that are used by highly skilled operators to machine sophisticated parts, made to finest tolerances. David Noble explains how numerical control (N/C) replaces these operators.

> Here the motions of the machine tool required to produce a particular part were described in detail mathematically, corresponding to the blueprint specifications for the part, and were recorded as numerical information, coded for economy, on the storage medium. The entire process of producing a part, including the skill of the machinist, was reduced to formal, abstract description, coded, and then translated (usually by a computer) into fully interpolated data to actuate the machine controls. Whereas with the motional approach, the skills and tacit knowledge of the machinist were automatically recorded as he interpreted the blueprint and put the machine through its

> paces manually, without ever having to be formally or explicitly articulated, with N/C, all interpretation was performed by a "part programmer," at his office desk, who was required to spell out precisely in mathematical and algorithmic terms what had heretofore been largely sight, sound and feel. Whereas record-playback was a reproducer and, thus, a multiplier of skill, extending the reach of the machinist, N/C was an abstract synthesizer of skill, circumventing and eliminating altogether the need for the machinist. In short, as one early N/C inventor described it, N/C was an "automatic machinist."[40]

Noble's major concern is to expose the social and political background of this development. Without intending to belittle it, I want to call attention to the losses of community and engagement that have not escaped Noble either. He speaks of the machinist's skill as "the fundamental and irreplaceable store of the inherited intelligence of metal working production."[41] But with the advent of automation, the feel, the intuition, the judgment, and the experience of the worker, handed down from master to apprentice, renewed and alive in each generation of workers, are lost. The community of workers, a republic of skilled and relatively independent artisans, is dissolved. In an appendix, Noble cites Kurt Vonnegut's poignant contrast of the worker and the technological text that has replaced him.

> And here, now, this little loop in the box before Paul, here was Rudy as Rudy had been to his machine that afternoon—Rudy, the turner-on of power, the setter of speeds, the controller of the cutting tool. This was the essence of Rudy as far as his machine was concerned, as far as the economy was concerned . . . . The tape was the essence distilled from the small, polite man with the big hands and black fingernails; from the man who thought the world could be saved if everyone read a verse from the Bible every night; . . . . Now, by switching in lathes on a master panel and feeding them signals from the tape, Paul could make the essence of Rudy Hertz produce one, ten, a hundred, or a thousand of the shafts.[42]

Let me now attempt a broader view of instrumental texts and suggest a connection to presentational texts. There is of course an endless variety of instrumental texts along with many different degrees of self-realization. As of now, most instrumental texts realize themselves less than perfectly. Their self-realization typically requires the skill and engagement of a human being to be useful. The tendency is to link instrumental texts with one another and to exclude the human members of the chain. The telos is one instrumental text whose self-realization is, e.g., the entire operation of a factory including not only production but also scheduling, inventory, shipping, billing, accounting, etc. Automation is proceeding haltingly and with lower productivity gains than expected. But it remains the object of enormous energy and ingenuity. In science fiction one can imagine societies where the various instrumental texts have spread and coalesced into a fully self-realizing supertext, leaving no gaps at all for humans in the gigantic process of production and administration. In actuality, this state of affairs is a regulative idea, nothing less or more.

Perfect and all-encompassing self-realization is the formal telos of instrumental texts. What is their substantive goal? Noble gives a succinct statement of the task to which the practitioners of science and technology, the authors of the technological texts, have officially dedicated themselves to. Their claim is

> that inevitably their combined understanding and know-how would enlarge society's supply of goods and services, lessen the burden of human toil, reduce pain, increase comfort, and expand the horizons of human freedom.[43]

This is a standard version of the founding promise of technology. It extends the traditional promise of prosperity and liberty, and, as is also usual, it conceals the dubious way this promise has been acted out with grand ambiguity. Not that the promise has been empty. By way of technology, the advanced industrial societies have greatly advanced against hunger, disease, and illiteracy.

But these gains are themselves ambiguous since in the abstract they could be the basis for very different kinds of positive freedom and wealth. In fact, however, this ambiguity has been resolved with growing definiteness. The pursuit of liberty has become the goal of disburdenment. In the terms of this essay, *to be disburdened* is to be rid of the duress and demands of reality. These burdens are to be shifted to self-realizing texts.

This sort of liberty still leaves us with the question of how we are to prosper in the space that has been cleared by the technology of disburdenment without again being restricted and imposed upon. The answer has been given succinctly by the Chairman of the Board of Microsoft Corporation. "In twenty years the Information Age will be here, absolutely. The dream of having the world database at your fingertips will have become a reality." And more specifically he says: "The world will be online, and we'll be able to simulate just about anything."[44] In our terms, prosperity will be the availability of indefinitely many and different self-realizing presentational texts. We will, each one of us, surround ourselves with a private sphere of pure appearances. No competence is required to do so, nor is another person, far less a community. Reality will be reduced to the staging area where we learn the rudiments of life that one has to command to recognize and appreciate the luxuriant appearances of real life. And reality will be exploited to yield the material and motifs for the construction of the appearances.

Even the electronically constituted world, resting on technological texts, will need a living community for its continued existence. But all it requires is a small community of technical experts. Literacy can shrink to a small fraction of the population. This is compatible with the survival of formal democracy. But the latter would lose all substance. This development is not inevitable. Culture and electronic technology as of now and as a matter of fact are fused into one tendency. But it is possible to recognize the pattern of this fusion and to arrest and redirect the tendency. It is possible to recognize instrumentalities and appearances for what they are and to have them serve rather than dissolve community and reality.

Presentational texts will do so if they serve education, science, or art. The elaborate and astounding presentational texts of flight simulation serve to train pilots.[45] Medical imaging aids medical science to explore the living human organism in three dimensions without mechanical intrusion.[46] And some of the presentational texts that we know as films surely constitute works of art that illuminate and help us to orient ourselves in the real world. But all these helpful presentational texts need the balance and warrant of the communal and engaging life that we are enabled to live through the deep realization of great traditional texts.

Endnotes

[1]See Walter J. Ong, *Orality and Literacy* (London: Methuen, 1982) and Joshua Meyrowitz, *No Sense of Place* (New York: Oxford University Press, 1985).

[2]In Nelson Goodman's terms, traditional texts consist of scripts, scores, and architectural plans all of which are more or less in a notational scheme. See his *Languages of Art*, 2nd ed. (Indianapolis: Hackett, 1976).

[3]See Ong, pp. 32-33, 81, 91.

[4]See *The Works of Horace*, ed. E.C. Wickham, vol. 1 (Oxford: Clarendon, 1877) 1-2; Eduard Fraenkel, *Horace* (Oxford: Clarendon, 1957), p. 20.

[5]Wickham, pp. xiii-xxi.

[6]See Charles Amzi Vannoy, *Studies on the Athena Parthenos of Pheidias*, Ph.D. diss., State University of Idaho, 1914, pp. 54-55 in particular.

[7]Goodman would say that a sculpture is an autographic work of art, and copies do not preserve its authenticity. A poem is an allographic work of art, and a properly spelled copy is as authentic as the original. See pp. 99-123. See also John Haugeland, *Artificial Intelligence* (Cambridge, MA: MIT Press, 1985), p. 55.

[8]Haugeland, pp. 58-63.

[9]See Goodman, pp. 127-73, for an elaboration of these points.

[10]No grammar has come close to giving a formal account of the rules of composition for natural languages. See below for more.

[11]See Kristi Vaughan, "Crumbling Culture," *Yale Alumni Magazine* 50

(Summer 1987), pp. 34-38.

[12]Ong, pp. 5-77.

[13]Ong is aware of both phenomena but pursues neither. See pp. 33, 81, 135-38.

[14]See John Chadwick, *The Decipherment of Linear B* (Cambridge: Cambridge University Press, 1958).

[15]How, then, can an unknown script be deciphered? By discovering residual bonds of community and world appropriation that we still share with the authors of these scripts (e.g., knowledge of places and their names) and by extending and recovering our common understanding from there. See Chadwick and E.A. Wallis Budge, *The Rosetta Stone* (London: Trustees of the British Museum, 1913).

[16]See D.C. Mitchell, *The Process of Reading* (Chichester: Wiley, 1982), pp. 6-7.

[17]See Frank R. Vellutino, "Dyslexia," *Scientific American* 256 (March 1987), pp. 34-41.

[18]Mitchell, pp. 190-95.

[19]See Jeanne S. Chall, *Stages of Reading Development* (New York: McGraw, 1983).

[20]For an instructive failure to accomplish the impossible see Douglas Waples, Bernard Berelson, and Franklyn R. Bradshaw, *What Reading Does to People* (Chicago: University of Chicago Press, 1940).

[21]See Chall, pp. 13-24.

[22]See, e.g., Irving M. Copi, *Symbolic Logic*, 3rd ed. (New York: Macmillan, 1967), pp. 198-201.

[23]We can put these points in the terms of Lakoff and Johnson and say that symbolic logic is partially structured by metaphors. These metaphors are grounded in our bodily experiences with the tangible world; and every such metaphor highlights some feature and hides others of the phenomenon it structures. See George Lakoff and Mark Johnson, *Metaphors We Live By* (Chicago: University of Chicago Press, 1980) and Mark Johnson, *The Body in the Mind* (Chicago: University of Chicago Press, 1987).

[24]On the significance and openness of the context of mathematics, see Philip J. Davis and Reuben Hersh, *Descartes' Dream* (San Diego: Harcourt, 1986), pp. 194-95, 198-99.

[25]See Roy Harris, *The Origin of Writing* (LaSalle, IL: Open Court, 1986).

[26]As record keeping develops into literature, trust remains crucial. See Robert J. Dostal, "The World Never Lost: The Hermeneutics of Trust," *Philosophy and Phenomenological Research* 47 (1987), pp. 413-34.

[27]See Harris, pp. 16-17.

[28]Parsimony, however, is not an overriding consideration for writing. Hence alphabetic writing is not intrinsically superior to nonalphabetic kinds. See Harris.

[29]See Ong, pp. 78-138.

[30]This chimera is haunting Peter Kivy in "Platonism in Music: A Kind of Defense," *Grazer philosophische Studien* 19 (1983), pp. 109-29 and in "Platonism in Music: Another Kind of Defense," *American Philosophical Quarterly* 24 (1987), pp. 245-52.

[31]See Johnson.

[32]See Norman Maclean, *A River Runs Through It and Other Stories* (Chicago: University of Chicago Press, 1976).

[33]See Nikolaus Harnoncourt, "Johann Sebastian Bach Mass in B Minor, BWV 232," liner notes to the Telefunken recording of April/May, 1968, SKH 20.

[34]Another necessary condition is the existence of canonical texts. We should search for an inclusive canon, avoiding male and Western chauvinism.

[35]See John Monforte, "The Digital Reproduction of Sound," *Scientific American* 251 (Dec. 1984), pp. 78-83.

[36]See Max V. Mathews and John R. Pierce, "The Computer as a Musical Instrument," *Scientific American* 256 (Feb. 1987): 126-33.

[37]See Stewart Brand, *The Media Lab* (New York: Viking, 1987).

[38]See "Mad About M-M-Max," *Newsweek* 20 (April 1987), pp. 58-64.

[39]See Marion Long, "The Seers' Catalog," *Omni* 9 (Jan. 1987), p. 38.

[40]David E. Noble, *Forces of Production* (New York: Oxford University Press, 1984), p. 84.

[41]Noble, p. 150.

[42]Kurt Vonnegut, *Player Piano* (New York: Avon Books, 1967), p. 38. See also Noble, pp. 359-60. Advertising, in its disarming and ruthless candor, proclaims the wonders and benefits of one of the offshoots of this development.

> NOW ONE OF YOUR MECHANIC'S TOOLS IS ANOTHER MAN'S MIND.
>
> At GM, we think of computers in human terms. Because today, we have begun to program human knowledge and logic into a computer. It's called artificial intelligence.
> So even if an engine expert retires from GM, his mind can still work for you. His lifetime of experience can go into a computer. The computer can then

dispense this invaluable knowledge not only to GM engineers, but to your mechanic, too. That way, he'll be able to use his computer to help him service your car.

With artificial intelligence, a lifetime of experience can now last a hundred lifetimes. Because at GM, our quest for knowledge never tires. Or retires.

In Time Campus 6 (April 1987), p. 5.

[43]Noble, p. 42.

[44]Long, p. 38.

[45]Ralph Norman Haber, "Flight Simulation," *Scientific American* 255 (July 1986), pp. 96-103.

[46]See Brand, p. 90.

LITERACY, MEDIACY AND TECHNOLOGICAL DETERMINISM*

Larry Hickman

Abstract: Those whom I designate "text-type determinists," including Albert Borgmann, grade texts into paper-based "traditional" ones, which they claim to be superior because their use makes us critical, and mylar-, celluloid- and vinyl- based "technological" ones, which they claim to be inferior because their use makes us lazy. Their argument is essentialist and determinist: it claims that texts have essences that determine the form of life of their users. Against them I advance the instrumentalist argument that texts have essences only in the functional sense that they comprise sets of constraints and sets of facilities. Regardless of their type, texts are tools which may be used to enhance delight and resolve difficulties; sorting them into inferior and superior types should be done on the basis of the extent to which this is possible. The development and use of critical intelligence is not uniquely linked to any one text-type. To be "mediate" is to engage texts of all types, including traditional literary ones.

Those of us who make it our business to think about the nature and function of what we call "texts" exhibit a notorious lack of consensus regarding just what should be counted as a text. There is deep disagreement concerning the partition of texts into types, as well as concerning the identification of text-tokens within text-types. Further, as

---

Lifeworld and Technology, eds. Timothy Casey and Lester Embree, Washington, D.C.: The Center for Advanced Research in Phenomenology and University Press of America, 1989.

Joseph Margolis has reminded us, talk about references to texts often gives way to talk about "textuality" and "referentiality."[1]

For some, a text is a literary text. Period. Others, among whom some are philosophers, have invited us to share their view that there exists an essential difference between paper-based texts and those that are produced or reproduced on other media. Albert Borgmann has constructed just such a textiary.[2] The type he calls "traditional" texts include literary texts, printed musical scores and architectural drawings. The type he calls "technological texts" include information on tapes, disks, microelectronic chips and celluloid.

I find Borgmann's choice of terms quite curious. The printing press in its many forms is also a form of technology, and so printed pages are also technological texts. For that matter, handwritten manuscripts are technological texts. The same could be said of stories told in languages that have no written form and of cave paintings: they too are "technological" texts because they are artifacts whose production involves the use of other artifacts.

Borgmann wants not only to separate the world of texts into two hemispheres, the traditional and the technological, but also to claim that one hemisphere is superior to the other. He tells us that "traditional" texts are superior because they are permanent, self contained objects that are closed off and distanced from reality, that their rigor is guaranteed by their immutable identity and their pure form, and that they enforce certain community experiences and practices that are salutary.

"Technological" texts are said by Borgmann to be inferior because they tend to make us lazy. They give us so much information that they require little reconstructive participation. Their realization and appropriation does not demand of us the effort required by literary texts. They divorce us from reality and create in us responses which prevent our participation in a community. This philosopher also writes that "reading appears to provide a richer and more enduring wellspring of conversation than . . . a film."[3] In his view, the "communal virtues of literacy . . . are . . . mortally threatened by the electronic media."

Since it seems reasonable to assume that readers of this essay spend much of their time developing their own literacy and urging their students to do likewise, perhaps Borgmann's manner of demarcating texts and the subsequent order of superiority and inferiority he accords them will find ready acceptance. Is it not after all true that a great obstacle to the careful reading of philosophical texts by our students is the time they spend with "the media"– meaning the *other*, non-print media? Aren't those media the sources of their confusion of psychological and biographical causes with logical reasons? Isn't it from those media that they get their their subjectivism, their idea that one opinion is as good as another, that we are all, as Pirandello suggested, right just in case we think we are?

Borgmann's company of fellow travelers appears to be a sizeable one. His dark warning concerning our students' preoccupation with non-print texts has been echoed by the writer of a current bestseller. Allan Bloom, the thesis of whose title appears, ironically, to be supported by his sales, has assured us that it is impossible for students to know how to view films unless they have first read Dostoevsky and Dickens. I quote from *The Closing of the American Mind.*

> As films have emancipated themselves from the literary tyranny under which they suffered and which gave them a bad conscience, the ones with serious pretensions have become intolerably ignorant and manipulative. The distance from the contemporary and its high seriousness that students most need in order not to indulge their petty desires and to discover what is most serious about themselves cannot be found in the cinema, which now only knows the present.[4]

Perhaps these somber warnings are correct. Perhaps non-print mediacy, the reading of "technological," or non-print texts, crowds out literacy, the reading of "traditional" ones.

Confident rejection of new text-types is certainly not novel in the works of Borgmann and Bloom. In his *Phaedrus* Plato had the Egyptian king Ammon warn Thoth, the inventor of writing, that his newfangled papyrus-based information retrieval system would prove a disaster. If the young and inexperienced began to occupy their spare hours with reading and writing, they would no longer avail themselves of the rich tutelage of those who were older and wiser. Learning would be more accessible—more democratic—but thinner and less personal. The realities comprehended by the oral texts would be lost, and there would follow dissolution of the community. The world would then comprise people filled not with wisdom, but with the conceit of wisdom. Literacy would crowd out memory, and democracy would crowd out authority. Civilization as Ammon knew it would be at an end.

One of the principal difficulties with the contemporary version of Ammon's view, the version held by both Bloom and Borgmann, is that it constitutes a species of technological determinism[5] which, like others of its genus, ignores certain facts. I shall call the Bloom-Borgmann thesis "text-type determinism." It claims that once a type of text is utilized it shapes us in ways that are beyond our control; that on balance, what our texts *are*, and consequently what they do *to* us, is of greater significance than what we do *with* our texts. Texts of the print-type are said to have the essential feature that once utilized they make of us critical and intelligent people. Texts of the cinematic-, photographic- and audio-type are said to have essential features that are different: their use is said to make of us people who are less intelligent and less critical than we would be if we did not use them.

The claim of text-type determinists thus differs from another, undeniable one that the kinds of instruments we utilize, in this case our choice of texts, constitute a form of life, a fabric of habits and working hypotheses by means of which we enter forearmed into new and problematic situations. This other view differs from text-type determinism in its assertion that it is the task of intelligence, regardless of the type of text to which it gives rise or to which it is applied, to improve hypotheses and to reform habits.

Their claim also differs from the one that says that we grasp and utilize different types of tools, hence texts, in different ways. That I handle a hammer in a different manner than I would a saw is indisputable. The enormous variety of tools and implements available for use in complex civilizations does not validate technological determinism and its concomitant essentialism so much as it attests to the prodigious adaptability and inventiveness exhibited by human beings.

The claim of the text-type determinists is much stronger than either of these views: it holds that once chosen, mechanical and electronic production and reproduction of images and sounds (as opposed to mechanical and electronic reproduction of marks on paper) decivilizes and renders passive those who involve themselves with such texts. Their claim is that non-print mediacy, the use of non-print texts, casts out literacy because if we are mediate we cannot be literate. They tell us that non-print mediacy involves the loss of critical skills, including the ability to interact constructively with literary texts. Their claim holds that text-types are uniquely linked to certain forms of life, and that this is true because each type of text has a unique essence and because the acceptance and use of one text-type in lieu of another closes out certain parts of our future.

It may be objected that I have been unfair. Perhaps the view I am opposing asserts no more than that people spend too much time going to the movies and watching television and listening to recorded music, and not enough time reading books and writing essays and making their own music. But the relative amount of time spent doing these things is not the central concern of the text-type determinists.[6] What alarms them is that people go to films or watch television or listen to recorded music *at all*. It is not just that such activities take the time which would otherwise be spent reading, but rather that they make us into different and lesser people than we would have been if we had not engaged in them.

Against this deterministic view I offer an alternative: I suggest that texts, whether printed, on celluloid, on mylar, or on photographic paper, or for that matter carved in stone, are nothing more or less than tools for use and enjoyment. Further, I suggest that energy spent attempting to find ways of solving perceived problems by experimenting with existing and proposed tools—finding out what we can do with them—is far more productive of augmented critical skills than attempting to discover their essences, their pure forms, their "inherent" qualities. To put this suggestion another way, what our tools do to and for us is dependent upon choices we continue to make, dependent upon our relations to them, not upon their "essence" or "pure form."

Text-type determinists contend that our tools have fixed essences and that we are what we are because of what they, our chosen tools, are. To this I would answer that ontological inquiries into the essence of our tools may possess certain aesthetic charms, but that such inquiries have notoriously failed to provide guides for action. One of the dangers of this ontologizing of texts is that it prevents meaningful inquiry into what they can be.

It was not clear, for example, during the first hundred years of printing what could be done with literary texts. The invention of the novel, to take an obvious example, was yet to come. And those readers in the years that immediately preceded the publication of *Ulysses* and *Finnegan's Wake* who thought that they had ascertained the "pure form" or "immutable identity" of literary texts must have been greatly surprised at what James Joyce wrought. Given the history of literary texts, it would be arrogant indeed to say that the essence of that text-type has now been, or ever will be, completely revealed.

Even if today it were known with certainty that we have once and for all time run out of genres, types and approaches to literary texts, that the well of invention will be henceforth and forever dry, the argument of the determinists that the essences of literary texts are secure would still be far from convincing. The texts of Shakespeare, Dickens and Woolf still provide novel insights even to, and perhaps especially for, their most accomplished readers. Discovery of novel uses of printed texts—choices we can and do make about their use—runs backwards into and reconstructs the past just as surely as it looks forward to and influences the future.

To say of a whole group of texts, then, to say of electronically and mechanically produced and reproduced images and sounds, that their essence has been grasped and that they can be nothing more than what they now are, or that their use denotes and promotes lack of critical intelligence, seems to me to fly in the face of well established fact. Text-determinism is a thesis that competes for lack of sophistication with Marx's claim in *The Poverty of Philosophy* that the hand mill begets a feudal society, and the steam mill begets capitalism.

The view advanced by these determinists ignores the work of video artists such as Peter Campus who have richly informed our vision. It neglects the work of filmmakers such as François Truffaut who have expanded the boundaries of what it means to construct a moving image. It demeans the work of photographers such as August Sander who have provided us with the means to look at the world in new ways. And most fundamentally it belittles the audiences which these artists have created, those individuals who work *with* and along side these artists in libraries, in their homes, in museums, and in darkened theatres, to reconstruct those texts for their own needs and purposes.

Rejection of text-determinism need not commit one to any type of subjectivism or relativism. The bald, indisputable fact is that human beings are in constant communication with one another. We communicate intentionally and unintentionally. Those long dead still communicate to us. Nor is great spatial distance any longer a barrier to communication. It is only by means of tools, including printed texts, photographs, films and videotapes, that we are able to do this. The photographer Russell Lee still speaks to us eloquently of the life of the dust-bowl farmers of the Great Depression, just as do the recording artist Woody Guthrie, the filmmaker John Ford and the novelist John Steinbeck.

The sources and types of significant texts are many and varied. Like other types of tools, some texts, perhaps most, are inherited. Some are found lying about and some are available from stocks that are common. Some are available only from sources that cater to the specialist. Some are borrowed, some are modified for personal use from stock items, and some are invented. We discard some texts as no longer useful and we use some texts only once and then replace them with another of their type. Some projected texts never leave the planning stages.

I ask you to notice that I have made no reference to inner and outer, subjective and objective, real and ideal in this talk of texts as tools. Like other types of tools, a text is an artifact that is produced for use and enjoyment. It may be a simple text designed for a specific task, such as a list of the current prices of mutual funds or a snapshot or videotape of a family outing. It may be a complex text capable of multiform usage and such as to give great delight, as do certain novels of Dickens and certain films by Truffaut. Good texts, texts that serve us well, are constructed so as to take into account the environing conditions that occasion their construction, use and enjoyment. Poor texts are not well constructed and so their use and enjoyment is flawed.

That there is much that is dreary on television and on film and on audio recordings is beyond dispute. There is also much that is dreary in print. That commercial television is used by some as a soporific only a fool would wish to deny. But it is also beyond denial that there are books–even some written by philosophers–that serve the same function.

What I urge, then, is that we put an end to speaking of tools, and therefore of texts, as having complete essences which predetermine and provide the measure of our ways of involvement with them. I suggest that we instead speak of the ways in which texts can and do serve to enhance delight and to resolve problems.

An important corollary to this view is the thesis that critical intelligence is not split into higher and lower realms along lines of division between involvement with texts that are "traditional" and texts that are "technological." What our text-type determinists argue is that films and photographs and videotapes are incapable of being viewed critically, that there is something in their essence that prohibits critical involvement with them. It seems to me rather the case that if our students do not know how to view a film, or how to look at a photograph, then it is because they have not been taught how to do so. If they are manipulated by images used for the purposes of commercial or political propaganda, then it is because they have not been trained to view images with care and insight. And if they do not demand high standards for the images and sounds that are produced by our culture, then it is not because they spend too much time with non-print media but because they are not sufficiently "mediate." To be mediate, as I use the term, is to have a critical understanding of the ways in which texts do and can work–both within and across text-types.

This is precisely the debate that Walter Benjamin joined in his 1936 essay "The Work of Art in the Age of Mechanical Reproduction."[7] A part of Benjamin's purpose in that essay was an assault on the cult of high culture, the equation of rareness with significance. Though this cult still exhibits vital signs, Benjamin was, I suggest, substantially correct when he argued that it was a symptom of an unfortunate and anachronistic social arrangement. But Benjamin also argued that there is an act of reconstruction, of complicity, in the way we deal with mechanically reproduced images. In looking at a film, says Benjamin, we identify not with the actor, but with the camera. Viewing a film is montage in reverse.

In his recent novel *Three Farmers on their Way to A Dance*, Richard Powers has provided this felicitous paraphrase of Benjamin's point.

> The idea that we look at the . . . subject matter begins to lose its credibility. We look over it, attempting to locate something else. When a movie editor cuts from a woman turning her head to a midrange shot of the shops across the street, we follow her gaze, having of our own volition turned our attention to look with her. To make sense of a montage means to reassemble the cut, in reverse, using the editor's criteria. We remake the montage dynamically, as an act of looking.
>
> The same applies to still photos: the lens slices a cross-section through time, presenting an unchanging porthole on a changed event. The frame invites us to feel a synchroneity with the photographer the way museum-case glass, slicing through a beehive, invites us to live in the colony . . . If the path

> between sense and significance opens, it will open in those moments when, momentarily delighted by some overlooked detail or construed resemblance, we become aware of what lies in front of the plane of the photo.[8]

What Benjamin and Powers have striven to tell us is that there is work that is constitutive, work that is productive, going on not only behind the lens but also in front of the surfaces of the photograph, the film screen, and the cathode ray tube. There is no simple act of looking; the eye is a prehensile organ that handles and feels and grasps its subject, turning it over and manipulating it to get a clearer view not only of the image itself, but of possible contexts from which it has been abstracted.

Powers' metaphor of the photographic image as cross-section is itself a productive one, for it reconnects two kinds of activities between which the text-determinists have attempted to drive a wedge. The activities which give rise to the creation and appreciation of images are reunited with those of the type we call "knowing" in a broader sense.

Powers treats the cross-section as a kind of hypothesis. It is a grasp put forth within a field of experience, a freezing of action, an arrest of movement. It is a location of sharpened focus. Productive knowing utilizes such specimens in order to get a firmer grasp on the broader field from which they are abstracted. Cross-sections/hypotheses allow experimentation with experienced fields; they help to indicate the possibilities of what experienced situations can become as a result of productive intervention into them. To look for the pure form or the essence of the cross-section is to miss the point of its function. A cross-section, whether a printed essay, an idea or a photographic image, is a proposal for something to be done.

Put another way, both creation and appreciation, whether of ideas or of photographic, cinematic, or video images, involve the bringing to a provisional close of a creative activity. Stopping of action is to some purpose: this, rather than that, is selected from a busy field. But the stop is provisional because it is the occasion for evaluation, for the positioning of the image or idea in ways that allow it to be utilized in the future with respect to things we can and should do, as well as with respect to cross-sections that are not yet had. With images, as with ideas, creation and appreciation broaden the field of actual and possible connections, and they heighten critical awareness.

The text-type determinism advanced by both Borgmann and Bloom is a part of a long and still powerful tradition which stands opposed to this view of the unity of intelligence and the critical activity of vision. Descartes spoke for this tradition when in Rule 12 of the *Rules for the Direction of the Mind* he treated vision as a passive conduit which is sensate only when being *acted upon*. His metaphor was still a passive one when he turned in the third of his *Meditations* to visual metaphors, clarity and distinctness, as criteria of truth. For this tradition, sight operates in a fashion that may be called "linear" insofar as it establishes a point to point conductance of information from the pure form of a complete and finished object to the visual orifice and thence to the common sense of the mind.

Against this view Benjamin argued that vision is "non-linear," that it is not passive and inert but actively exploratory. This alternative account of vision constitutes a radical departure from Descartes' metaphor of rays entering the eye from a finished object. In this alternative account the eye is a probing tool that moves and checks and interacts. Sidney Hook drew attention to this constructive activity of vision in 1927 in *The Metaphysics of Pragmatism*. "We use our eyes," he wrote, "as we do our hands–to grope, to pry, to scan, to escape danger, to signal a friend. Sense activity, like all behavior generally, is not inertly receptive in the presence of stimuli or explosively active in their absence. It is *interactive.*"[10]

***

There are two great fallacies involved in technological determinism, hence in the species I have called text-determinism. The first is a reductive fallacy which says "artifacts of the type A have effect y, therefore no other type of artifact can have effect y." Bloom's comment on Dickens offers a clear example of this fallacy. ". . . students today have nothing like the Dickens who gave so many of us the unforgettable Pecksniffs, Micawbers, Pips, with which we sharpened our vision, allowing us some subtlety in our distinction of human types. It is a complex set of experiences that enables us to say so simply, 'He is a Scrooge.' Without literature, no such observations are possible and the fine art of comparison is lost."[11] Bloom tells us in this passage that literary texts and no other kind of texts allow us to sharpen our vision with respect to human foibles and character flaws. It is his view that such insights are not attainable as a result of interaction with any other type of text. We cannot expand our understanding of human behavior by viewing films critically.

It is doubtless true that there are some novels that should never be translated into film. E. M. Forster said that his *Passage to India* was one of those novels, and I for one think it unfortunate that David Lean did not take him at his word. Further, I take it as obvious that there is a *prima facie* richness about a novel due to the time required to read it that is difficult to approach in the two or fewer hours it takes to view a film. But the same may be said of the text-type we call the short story. To say this is just to admit the obvious, that we use the tools we call texts in different ways, and that there is no perfect translation from one medium to another. Films based on books offer enormous problems of translatability; but so do books–and essays–about films.

To say, however, as do the text-determinists, that we get a more detailed impression of the poisoned political prisoner in Manuel Puig's novel *Kiss of the Spider Woman* than we do in Hector Babenko's excellent film version of that novel is nonsense. Words on a page cannot convey the convulsions of a poison victim with the same intensity that is managed by a moving image in the hands of an accomplished film director. To take a case of political documentary, no assemblage of print-texts was able to galvanize a nation against the indignities of racial segregation as did the film footage of Bull Connors loosing his dogs on demonstrators in Selma, Alabama. If Borgmann wishes to speak of media in terms of their abilities to provide wellsprings for conversation, then I submit that film footage as an example of a text of enormous consequence.

There are those who will object that images, because they can carry heavy emotional charges, are capable of political abuse. In the hands of the wrong people images can become instruments of repression and destruction. Another form of this objection is the claim that filmmakers sometimes serve purposes that are evil even though their own motivations, as those of the director of *The Triumph of the Will* arguably were, are more innocently aesthetic. Thus put, what has been said is unquestionably true. But the text-type determinist wishes to go further to say that cinematic and video and photographic texts are capable of manipulation in ways that print texts are not because, unlike print texts, they are not amenable to critical interpretation. This is precisely the reduction advanced by Bloom.

Unfortunately, however, his view ignores several important facts. It ignores the fact that print texts have also functioned as means of manipulation, as did Stowe's *Uncle Tom's Cabin*. It ignores the fact that print texts can and do function as instruments for critical assessment of non-print texts, as has Eisenstein's *Film Form*. And it ignores the fact that communities of inquiry into non-print texts, such as film societies, have grown up around their respective texts in the same way that other communities of inquiry have focussed upon print texts.

Even more significantly, however, this type of reductionism yields the field to those who create commercial and political propaganda. It denies that intelligent estimation of image-texts or music-texts is possible; it denies that education in this field can be effective.

The remarks of Sergei Eisenstein, who was also interested in the works of Dickens, provide a kind of vaccine against the reductionist fallacy of Bloom and Borgmann. Eisenstein argued that cinema and literature have features which they share, as well as certain features which are unique to them. His remarks should not be taken in an ontological sense, however, but in terms of the ways in which cinema and literature can be utilized. In a passage that constitutes a sharp contrast to Bloom's, he writes:

> Perhaps the secret lies in Dickens' (as wellas cinema's) creation of an extraordinary plasticity. The observation in [Dickens'] novels is extraordinary as is their optical quality. The characters of Dickens are rounded with means as plastic and slightly exaggerated as are the screen heroes of today . . . It is absolutely thus that Dickens draws his characters—this is the faultlessly plastically grasped and pitilessly sharply sketched gallery of immortal Pickwicks, Dombeys, Fagins, Tackletons, and others . . .[12]

A bit later in the same essay Eisenstein chides as "thoughtless and presumptuous" those who think that they "can erect laws and an esthetic for cinema . . ."[13]

Eisenstein's point is that literary texts share many structures with the texts of other media, and that some of the same critical tools by means of which we operate in one arena quite often serve us well in another. He invites us to read Dickens from the standpoint of cinema mediacy, as well as to read cinema texts from the standpoint of the structure of Dickens' novels.

The second great fallacy of the text-type determinists is what John Dewey called "the philosophic fallacy." This is the mistake of taking some result of inquiry, hypostatizing it, and then positing it as something fixed and finished prior to inquiry, treating it as if it had been there all along. The text-type determinists conclude that non-print texts are for the most part treated uncritically in our culture. No impartial observer would deny them that conclusion. But then they turn their conclusion into something that is said to be an essential feature of non-print texts. They hypostatize what they have found in inquiry into something that preexisted it and that is said to have absolute existence in its own right.

The text-type determinists call attention to the widespread use in American culture of image-texts and music-texts. They also point to an equally evident fact, that ours is a society that has not come to terms with its educational needs, a society that has not developed meaningful patterns of leisure-time use. They conclude that the use of non-print texts constitutes a determining factor for de-education and laziness. In the view of Borgmann and Bloom, that is the essence of non-print texts.

But they have put the cart before the horse. It is not that we have text-types with essential structures at work determining human behavior, but rather that poor decisions have so limited our perspective that full inquiry into the possibilities of non-print texts has been lacking. One can admit that television-texts and cinema-texts are sometimes silly and sometimes even downright dangerous—and at the same time reject the determinist thesis that those text-types are essentially flawed and beyond redemption.

***

The view I have presented has no quarrel with those who say that there is much work to be done to improve the quality of most of the non-print texts utilized in our culture. There is also much that could be done to improve the use and enjoyment of print texts. What I have argued instead is that the development and use of critical intelligence is not uniquely linked to the use of one text-type; that we must engage with intelligence texts of all types, even if some of the text-tokens within those types are silly, dreary or even dangerous. To be mediate is to be literate, but it is to be more than literate.

Endnotes

* "This work was made possible in part through a grant from the Division of Research Programs of the National Endowment for the Humanities, grant number RH-20844-87."

[1]Joseph Margolis, "What is a Literary Text?" in *At The Boundaries: Proceedings of the Northeastern University Center for Literary Studies*, Vol. I, ed. Herbert L. Sussman (Boston: Northeastern University Press, 1983), pp. 47 ff.

[2]Albert Borgmann, "Texts and Things: Holding on to Reality," a paper delivered *Lifeworld and Technology*, a conference at Duquesne University, October, 1987. See also Borgmann's *Technology and the Character of Contemporary Life* (Chicago: University of Chicago Press, 1984).

[3]Borgamnn (1984), p. 34.

[4]Allan Bloom. *The Closing of the American Mind* (New York: Simon and Schuster, 1987), p. 64.

[5]Although the prospect is a tempting one, I cannot here provide a taxonomy of forms of technological determinism. One indication of the size of such a project, however, is that there appear to be at least as many varieties of technological determinism as there have been publishing Marxists: probably more, since Marx himself held several different varieties of the thesis during his lifetime.

[6]Bloom at least allows that listening to recorded *classical* music may not be antagonistic to literacy.

[7]Walter Benjamin, "The Work of Art in the Age of Mechanical Reproduction," in *Illuminations*, ed. Hannah Arendt (New York: Schocken Books, 1969), pp. 217-251.

[8]Richard Powers, *Three Farmers on Their Way to a Dance* (New York: McGraw-Hill, 1985), pp. 331-332.

[9]See John Dewey, *The Early Works*, ed. Jo Ann Boydston (Carbondale and Edwardsville: Southern Illinois University Press, 1969-1972), Vol. 3, p. 109.

[10]Sidney Hook, *The Metaphysics of Pragmatism* (Chicago: Open Court, 1927), p. 29.

[11]Bloom (1987), pp. 63-64.

[12]Sergei Eisenstein, "From 'Dickens, Griffith, and the Film Today,'" in *Film Theory and Criticism*, 3rd edition, ed. Gerald Mast and Marshall Cohen (New York: Oxford University Press, 1985), p. 371.

[13]*Ibid.*, p. 379.

MEDIA AS LIFEWORLD

Wolfgang Schirmacher

Abstract: Taking the example of music videos, I analyze the creating of an artificial lifeworld by this "new media." The modern perspectives of time, space, matter, language, person, and truth are employed as well as postmodern perspectives such as presencing, absence, hyperreality, difference, and happening. The foundation of media in communication as autobiography is exposing us to some of the future modes of living.

I. Media as Autobiography

When speaking or writing about media we still tend to think in terms of channels to which sender and receiver are connected, a temptingly simple but inadequate definition. Current media theory proposes the "Gutenberg Galaxy" has run its course, that the electronic media itself has become the "message" and its forms are to be understood as "extensions of man."[1] Following Marshall McLuhan and considering the stunning developments of the "new media" theorists such as Neil Postman, media is analyzed not only as a metaphor and representation of knowledge, but as an activity, one shaping our social environment and subject to a media ecology.[2] This shift in understanding necessitates a fundamental change in our relation to media, and in our thinking. Yet although the distinctions between reality and imagination, truth and fantasy seem to vanish, and the acceptance of media as an authentic lifeworld the next step, media theory is still reluctant

Lifeworld and Technology, eds. Timothy Casey and Lester Embree, Washington, D.C.: The Center for Advanced Research in Phenomenology and University Press of America, 1989.

to face this consequence. Signification, representation, the ideology of an independent reality as the measure of truth–these are compelling and long-held presuppositions not easily cast aside.

A phenomenological view, however, allows us to skip over the question of whether the world of the new media is real or not and instead describe *how* the new media shows itself to us. We then become aware of the seductive intensity, the speedy flow, and the open audio-visual textuality with which music videos, for example, "blur previously distinct separations and boundaries, such as those between popular and avant-garde art, between different genres and artistic modes, between past, present, and future."[3] Artistic media have long been considered as possessing a reality of their own, and the artist's work is often viewed as autobiography. In the new media of cinema[4] the trend toward self-portrayal is perhaps most obvious. Phenomenologically, such 'writing' of one's life in media reveals its structure as an *in-between* which is neither subjective nor objective. It fulfills an intentionality which transforms 'objective' material and 'subjective' goals into a living process. A lifeworld is commonly "lived through"[5] silently, the need to acknowledge is largely unfelt; in the case of mass media however only we can be held responsible for its superficiality. The model of nature with its self-evident presence is replaced by the model of technology which has to constantly give evidence and defend its presence anew.

In the technological age we can not simply live our lives, we have to write–and rewrite–them, or others will do it for us. Having been changed into this autobiographical 'writing,' communication can now be defined as authentic, as a responsible style of media. The difference between the "life lived through" and the world of media is still a sharp one in terms of perception, as Maurice Merleau-Ponty indicates,[6] but is an abstraction in terms of intersubjectivity as the basis of communication. In a lifeworld shaped by technology we consider as humane that for which we as artificial beings can and must take responsibility.[7] That society and mass media usually live our lives for us is entirely our own fault and not due to any flaw in the structure of these modes of our existence. We are, by nature, the artificial ones; rejecting our artificiality for a more 'natural' mode of being is a naive denial of our potential as well. Consequently our search should be for the authentic mode of artificiality. The artificial lifeworld created by us in the new media is of itself a fulfillment of humanness–a frightening outlook if one regards American prime time television as an example.

II. Music Videos: From Youth Culture to Artificial Life

McLuhan's pragmatic children use music videos as their "own comfy space" and as a "radio station for the eyes,"[8] enjoying the pleasure and magic of their youth culture. With the start of 24-hour cable MTV in 1981, music videos became the life style of an entire generation. Dweezil Zappa, the gifted son of a famous father "watches hours of MTV every day,"[9] and he is neither uneducated nor unemployed. MTV alone plays 75 - 110 clips a week, each with a life-span of 9 - 18 weeks; other channels (Black Entertainment Television, to name just one) add to this number. Performers such as Michael Jackson and Madonna have attained a personal iconographic power apparent on any city street.

In the eyes of one shocked admirer of Plato, the "ambition" of these young people seems to be "to win fame and wealth in imitating the drag queen who makes the music." Professor Bloom denounces music videos for turning life into "a nonstop, commercially prepackaged, masturbational fantasy."[10] And Kuan-Hsing Chan, an observer even more knowledgeable in video culture, shows deep concern when he writes: "If ecstasy of communication, fascination, desire, schizophrenic corruption of temporality and spatiality, obscenity (of sexuality), collage, quotation, fragmentation and non-unity are the key terms to describe . . . MTV, then we have moved from the question of 'what does it mean' to 'what does it do?'"[11]

Since music videos are at the core of today's youth culture, such harsh reaction to their seeming corruptive influence is predictable. Every youth culture antagonizes mainstream culture, criticizes the parent generation, and develops its own language and gestures, invents its own styles and idols. Undeterred by the intended shock effect of the new and unfamiliar, by the strangeness of certain appearances, a phenomenological approach will continue asking *how* the phenomenon of music videos shows itself to us, as long as it remains unknown. It should be noted that music videos are appreciated outside the youth culture as well. Station VH-1, for instance, is geared specifically to the adult audience. Music television is one of the few media hybrids interactively using other telecommunications media and live events for two-way communication in order to spark the imagination and interest of its viewers. These channels are continually re-evaluating and redefining themselves, and in so doing retain the vitality of the avant-garde.

III. Innovations in Communication: The Phenomenon of Music Video

Music videos are made for nearly every single released today and have become a major force in the music market. The leading channel, MTV, reaches 36 million people. Music videos have influenced, changed, colonized, cannibalized, and merged other types of media. And the viewer has become a collaborator. Music television seeks the constant feedback of its audience, altering its programming accordingly. Music videos seem to be a postmodern response to Richard Wagner's *Gesamtkunstwerk,* the attempt to unite the arts in the grand opera. Though music videos are only 'operetta,' they should not be underestimated: there is no question that this new medium is here to stay. Michael Jackson's recent 17-minute video "Bad," directed by none other than Martin Scorcese, is a clear indication that music videos are about to break out of their rigid format.

A phenomenological awareness is less interested in the content of a music video–decisive is rather the way it manifests itself. That many music videos represent "male street culture, social rebelliousness, and fantasies of female conquest,"[12] and that there is an answer from a "female culture" is not our concern here. Music videos are clearly an integral part of contemporary youth culture and for this reason alone an interesting field of research for both sociologists and semiologists. Nonetheless, our line of questioning will be the innovations in communication which music videos indicate apart from and beyond their role in youth culture, changes in the way we communicate with ourselves and with one another. This development is by no means restricted to music videos. But perhaps it is most visible in music videos–or perhaps they are simply the most fun to analyze! Their colorful and anarchistic texture, filled with shock images, their visual variety and ability to cut through clutter stimulates a sense of creativity too often missing in the technological lifeworld.

Before we consider the phenomenon of music videos in detail, I would like to stress the point that music videos belong to cinema and its aesthetics. The *movement-image* is, according to Deleuze, the basic unit of the cinematic discourse. This unit, a totality encompassing time and space not only makes the difference of opposing entities clear, it also allows them to be experienced as separate and distinct yet belonging to the whole. The movement-image, as a process, does not destroy the identity of the cinema, of movement and image. In the mode of unity the instant and the whole, immobility and change, self-awareness and a sense of the other are experienced in their discreteness as two differing and dialectically joined moments and, precisely through this relation, in their oneness. The cinematic movement-image gives us together and as one unit both the instant (i.e., the smallest perceptual unit of the shot) and the whole (which is our sense of change, flow, and duration).[13]

The music video is grounded in this a-centered unity of the movement-image which allows us a communication *in-between*. The characteristic modes of such communication are manifest in music videos and can be understood as creation of a lifeworld in which mediation becomes the reality. Our music video examples will include performance clips as well as concept clips. We shall consider two iconographic videos in which the performer is spotlighted: Prince's "You've Got the Look" and Michael Jackson's "Bad." You need to see only a bit to get the point; the details deserve our attention. In their shortness and compactness music videos reveal a structure often similar to that of a poem. Viewed in its entirety the concept clip "Dear God" from XTC offers a unique opportunity to find out how the problem of theodicy has changed since the time of Leibniz.

Let us examine music videos within the modern structural perspectives of a) time, b) space, c) matter, d) language, e) person, and f)truth in relation to their postmodern counterparts of presencing, rupture, hyperreality, writing, happening, and exposure.

1. From Time to Presencing

Time in music videos is created by cutting and montage. At first glance, performance clips seem to deal with real time, and concept clips to have more freedom in time disruption.[14] But *how* do music videos generate a sense of time? The rhythm, the pulsating beat, constitutive of most music videos, defines time. The driving, unrelenting beat, whether of sound and/or the visual beat marked by the rapidity of the image's movement, creates a time without past and future. The "reign of the presence"[15] is obvious and total. Even quotation of the past, as in Madonna's "Who's That Girl?," is merely material for an overwhelming universal coexistence, a time process of pure moments which could be called *presencing*.

The power of this presencing in music videos literally wipes out the sense of space. The stage character of the environment makes space appear artificial and interchangeable. The idea of progress which can create space has been abandoned in music videos. The beat, the gestures, the drive may still be rebellious but ultimately go nowhere. Prince's cruising and Michael Jackson's "loose-limbed prowess" demonstrate a stunning formalization of street rap, the art of movement without moving, a flowing with the beat in stillness. XTC's "Dear God" may seem to be narrative in composition and have a beginning and an end, but this impression is contradicted by the video's surrealistic elements. What we have is the presence of a nightmare without end and a narrative continuing on like a broken disc.

2. The Absence of Space: Rupture and Decentralization

A sense of space provides security and helps satisfy our desire to determine our place in the world. Merleau-Ponty points out that space is the distance between the body and the objects which create our own sphere.[16] Emotional distance allows us the experience of being the center. In music videos space is present in absence only, as rupture and decentralization. Immediate access to emotions and images destroys any sense of personal space. Behavioral as well as spatial scope have diminished. The "global village" realizes the end of distance; where barriers have fallen there can be no space. And the ruthless use of even the most expressive symbols has forgotten space. Rupture, experienced through distortion, need not be spectacular, as in Heavy Metal videos; it can be achieved through subtle changes, all the more effective. The video "Dear God" goes out of tune; this is effected by (a) varying shutter speed, (b) reversing tape direction, and (c) varying playback speed. Relevant to disorientation, Deleuze notes the viewer of cinema is acentered, lacking a fixed point of view. Technical manipulation in music videos and a "floating" viewer together create the rupture of personal space and a spreading out of presence. Temporality has become ever more important, incorporating all that which was formerly the realm of spatiality.

3. The Perceived World: Matter as Hyperreality

The slant of distortion is how music videos perceive our being-in-the-world, but this is not a pessimistic view. In fact, it reveals a multiperspectivity which gives us a more authentic world than the fake narratives of print media and television. The layerings in the beginning of Prince's video create, by the superimposition of different perspectives, a hyperreality which is, as Merleau-Ponty would say, more real than reality. Out of Michael Jackson's rumble dance on the stage of a New York subway platform in which film, music and dance are equally balanced, emerges a coherence in dissonance which is the material side of most music videos. Even "Dear God" is a realistic look-alike, only to turn surrealistic after a few moments.

4. Writing the Differences of Languages

Music videos can be viewed as an example of Jacques Derrida's criticism of our traditional logocentrism with its predominance of verbal language. The lyrics are composed of jokes and puns, intended to be self-ridiculing and confusing. Difference, not identity, is at stake. The equality of the verbal, visual, audio, and body language breaks the spell of meaning and reveals the whole of language which is beyond our instrumental grasp. Disconnection, distortion in pitch, and contradiction are on all levels the *movens/stimuli.*

Critics have attacked the non-narrative structure of music videos as impotent, unable to create an authentic lifeworld, and as an escape into the abstract. But the fact is that *STOP MAKING SENSE* creates a lifeworld in its own right experienced and shared by the millions who view music videos. Michael Jackson has been accused of "evading reality" and seeking an "occult solution,"[17] but his streetwise dialogue is meant to overcome a ghetto reality by stylization. This creates an artificial lifeworld in which communication is split and re-split in an *interchange* rather than bound to the meaning of words and to intersubjectivity. Derrida calls to mind that the writing of differences, the play of non-elimination, and changes in perspectives are needed in a lifeworld shaped by the violence of identity.[18]

Even the concept videos which have a message do not usually preach it, as we can see in "Dear God;" it is not a straightforward theodicy, but a tricky process of thinking *without abstraction*.

5. Icons, Persons, and Life Happening

Commonly understood, a person is defined as a certain life history, developing an identity in an intersubjective communication. Perhaps the most startling observation in music videos is that despite the personal iconographic power of the performers, the process of identification is *not* encouraged. This is not to say there is no strong business interest in repeating the hero identification attributed to youth culture. The popularity of the "Madonna Style" was notorious and visible throughout the USA.[19] But there are some significant differences worth noting. Icons such as David Bowie, Prince, Michael Jackson, attempt to diffuse their role model and obliterate (to some extent) race, sex, and age. Icons celebrate the bad and the ugly, scorning the dictates of a traditional aesthetic. Michael Jackson looks very much like a yelling woman! Involvement in a peer group becomes non-personal; it is no longer an interpersonal but an interfunctional relation, an interchange of performer, viewer, producer, engineer.

Music videos suggest a life as life happening, and the icons offer keys to this lifeworld of music videos much closer to art than to so-called reality, that which we conceive to be reality. Make it happen, let it happen!— Without history, without future, without present, only happenings exist.

6. The Exposure of Truth

The play of the revealed and the unrevealed (and the never to be revealed) is the play of truth as disclosure, according to Heidegger.[20] Music videos do not intend to be true, and in their elusive concreteness reveal what a direct approach would miss. When our interpretational models of meaning disappear, quite a different view of things opens up to us. Music videos expose a lifeworld which already exists but is still hidden from a knowledge which knows too much. Does this mean Michael Jackson is exposing the bad as fundamental, Prince the ambiguity of appearance, and XTC the absurdity of the human condition?

Much too simple. Music videos cannot be utilized for meaningful messages. The creative *how* is the "message." Music videos offer an exciting example of a concrete comprehension and a critical thinking of a kind which leads to discovery, not meaning. Amazement and wonder are not eliminated from this process; on the contrary, they are "traces" (Derrida) of and toward a world on the horizon of technology.

IV. Conclusions

Should mankind live to see its future–which is not very likely–music videos can be considered as the first hints of a lifeworld shaped by the conditions of a fully developed technology. We pride ourselves on our state-of-the-art technologies, blind to the actual primitiveness of these advancements. Present technology is still in its earliest developmental stages. The high technology of coming generations is a self-evident life technology functioning as naturally and imperceptibly as the techniques of breathing and walking. Outwardly, this technological world that is organized by a computer society, urban living, automated production, genetically engineered organisms, etc. will appear much the same as our world today. But the human attitude will be totally different. That which compels us to live in unrecognition, as instruments, that which makes our life inhuman, can by the same token develop our authentic human potential, given the opportunity to enfold within a non-instrumental manner of living.

Human beings will then be considered–as were the ancient Greeks–to be able to assume responsibility for their 'bios' and act free as artists, whatever their work may actually be. This life would be a truly artificial life and the fulfillment of humanity. This life style will be characterized by a wider range of life techniques which we can not yet imagine–but it is likely that music videos will expose to us some of the future modes of living. Like Moses, we shall never set foot in that promised land, but music videos, although only an accompanying phenomenon of contemporary life, give us a strobe-light glimpse of a lifeworld already here but not ours.

Endnotes

[1]Marshall McLuhan, *Understanding Media*, New York: Signet Books, 1964.

[2]Neil Postman, *Amusing Ourselves to Death*, New York: Viking Books, 1985.

[3]E. Ann Kaplan, "History, the Historical Spectator and Gender Address," in Music Television. Journal of Communication Inquiry, Winter, 1986 (10/1), p. 6.

[4]This term is used to describe the unity of film, television, music, photography, and video. Cf. Gilles Deleuze, *Cinema 1*, Minneapolis: University of Minnesota Press, 1986, chapter 4.

[5]Maurice Merleau-Ponty, *Phenomenology of Perception*, trans. Colin Smith, London: Routledge & Kegan Paul, 1962.

[6]Maurice Merleau-Ponty, "Film and the New Psychology," in *Sense and Non-Sense*, ed. Hubert Dreyfus, Evanston: Northwestern University Press, 1964, p. 48ff.

[7]Carl Mitcham, "Responsibility and Technology: The Expanding Relationship," in *Technology and Responsibility*, ed. Paul Durbin, Dordrecht: Reidel, 1987, p. 5-39; cf. Wolfgang Schirmacher, "Homo Generator: The Challenge of Gene Technology," in the same volume, p. 203-225.

[8]Barry Walters, "Like it or Not: MTV lives," in The Village Voice, June 2, 1987, p. 39.

[9]US, March 9, 1987, p. 54.

[10]Allan Bloom, *The Closing of the American Mind*, New York: Simon &

Schuster, 1987.

[11]Kaun-Hsing Cheng, "MTV: The (Dis)Appearance of Postmodern Semiosis, or the Cultural Politics of Resistance," in Journal of Communication Inquiry, Winter 1986 (10/1), p. 66.

[12]Lisa A. Lewis, "Female Address in Music Videos," in Journal of Communication Inquiry, Winter 1987 (11/1), p. 74.

[13]Cf. Deleuze, *Cinema 1*, op. cit.

[14]Cf. B. Walters, "Like it or Not: MTV lives," op. cit.

[15]David J. Tetzlaff, "MTV and the Politics of Postmodern Pop," in Journal of Communication Inquiry, Winter 1986 (10/1), p. 83.

[16]Cf. Merleau-Ponty, "Film and the New Psychology," op. cit.

[17]Stephan Holden, "The Dark Side of Peter Pan," in The New York Times, Sunday, Sept. 13, 1987 (Pop View).

[18]Jacques Derrida, *Margins of Philosophy*, ed. A. Bass, Chicago: The University of Chicago Press, 1982, p. 3ff.

[19]Lewis, op. cit., p. 80.

[20]Martin Heidegger, "The Turning," in Heidegger: *The Question Concerning Technology*, trans. William Lovitt, New York: Harper & Row, 1977, p. 36-49.

My special thanks to Carol Froehlich, Producer, BMG New York, for her assistance in the selection and editing of the music videos discussed in this paper.

A PERSPECTIVE ON THE RATIONALITY OF SCIENTIFIC TECHNOLOGY OR HOW TO BUY A CAR

Lester Embree

Abstract: Defined as the using of passive as well as active and nonartifactual as well as artifactual equipment, technology is recognized to have historical phases and stratified structural components that can be characterized as non-specialized, specialized, machine, and scientific. Once the constitution of equipment in the user's volitional life has been described via reflective observation, it is then related to the problematics of praxiological reason, which relates through axiotic and expistemic reason to the scientific and non-scientific evidencing of objects.

Scientific technology is changing the world at an accelerating rate. According to Lynn White Jr., "The emergence in widespread practice of the Baconian creed that scientific knowledge means technological power over nature can scarcely be dated before about 1850, save in the chemical industries, where it is anticipated in the eighteenth century. Its acceptance as the normal pattern of action may mark the greatest event in human history since the invention of agriculture, and perhaps in nonhuman terrestrial history as well."[1] Was there non-scientific technology prior to

Lifeworld and Technology, eds. Timothy Casey and Lester Embree, Washington, D.C.: The Center for Advanced Research in Phenomenology and University Press of America, 1989.

1850? Does it still exist? What makes technology scientific? Is scientific technology ipso facto the only rational technology? What is rationality? These and related questions are addressed in the following discussion. Examples are included for the sake of clarity, especially examples related to the using of the automobile, the most complex and powerful equipment familiar to most readers of this essay.

1. Technological Life

Technology bears much pondering. Considerations of history are especially valuable and readily show that so-called applied science is merely the most recent and most conspicuous part of an older and a greater whole. A general definition is needed. The word "technology" often merely signifies equipment today, but a modicum of reflection shows that things are equipment only in relation to how they are used. If "equipment" can then be used to signify equipment, "technology" can signify *the using of equipment*. Driving a car, for example, is using equipment, yet this characterization is essentially incomplete because equipment is always used for some purpose. Perhaps the car is used to get the driver to her workplace or to bring his groceries home from the market. As a rule, equipment is used to change some part of the world, i.e., to create, foster, or preserve or to destroy, impede, or prevent (whether there is neutral as well as such positive or negative total and partial purposes regarding objects that already exist or not does not need to be pursued here). "Equipment" denominates a practical object used as means to another practical object which is its end. Both practical objects are essential to technology. If the world is a vast totality of practical objects, then technology may be defined as *the using of equipment* on the world.

As used here, the signification of "equipment" is broadened in two major respects. First, not all equipment is artifactual. If one picks up a stone and flings it at prey, the stone becomes a missile but remains nonartifactual ("natural" is a word conveying too many significations). How much of the world remains unaffected by at least inadvertent human action is an interesting question and perhaps the best way in which to appreciate the vast extent of the artifactual is to search for nonartifacts in it. The sun and stars, which can become equipment in the technologies of navigation and astrology, are certainly nonartifactual.

Second, a distinction between *active* and *passive* equipment is necessary. There seems to be a tendency to think of equipment chiefly as items that are moved and guided like tools and cars. Lewis Mumford, however, also recognized "utensils, apparatus, and utilities. The basket and the pot stand for the first, the dye vat and the brick kiln stand for the second, and reservoirs and aqueducts and roads and buildings belong to the third class."[2] Utensils, apparatus, and utilities can be considered subspecies of *passive equipment* (Mumford does not, however, use "equipment" to express the here broadened signification nor write of "active" and "passive" species). If it is difficult at first to call utilities such as reservoirs and roads "equipment," at least they are plainly artifactual. But the forests in which one hunts, which are therefore hunting grounds, and the seas over which one sails, which then have sea routes, are both nonartifactual and passive equipment, if, that is, they are not affected by human action. Clearly, active equipment is active in relation to passive and vice versa and not only is equipment equipment only in relation to the use of it on the world, but it is also artifactual or not and active or passive in relation to such action.

Several comments are in order at this point. First, seemingly nontechnological activities, e.g., teaching, become technology insofar as they require at least passive equipment, e.g., classrooms, but the requisite equipment might not be specifically artifactual, as the case of Socrates and Meno's slave boy on an Athenian street shows, and if the teaching occurred out in the wild woods or on a beach, the teaching place would not need to be artifactual in the signification of "modified by human action" at all. In this case, the aspect of the world to be modified is the student's understanding. Second, one might think that this line of thinking leads to everything being technological. But equipment use pertains to a *practical attitude* in which some aspect of the course of the world is at issue, which is not the case in attitudes in which things are simply enjoyed or in which the goal is knowledge. And perhaps there are some activities that are utterly indifferent even where passive technology is concerned. Nevertheless, it seems wiser to begin with the technological and search for the nontechnological than vice versa. Third, the present perspective on technology can make the relations between philosophy of technology, on the one hand, and ethics and politics, on the other hand, problematic. Perhaps moral philosophy should no longer be modeled on social relations not mediated by equipment, as was the case in Plato and Aristotle.

Technology as use of equipment on the world can be classified in many ways, e.g., regarding the aspects of the world affected, the types of equipment used, and the skill and organization of the equipment users. The following sketch may be attractive because systematic questions about the relations among technologies of various sorts as well as events in the history of technology can be related to it.

(1) *Non-Specialized Technology*. Humans are not the only animals using equipment, but, as with communication, the complexity is vastly greater in humans. Nevertheless, comparison with brutes can be valuable. Among other things, "technology" connotes skill, which can be defined as learned ability. To use a stick to rake bananas into one's cage or to strip the leaves off a twig in order to fish dinner out of a termite hill does not require training, for they are techniques that can be learned through imitation of unwitting example. There seems a great deal of this in human life. "Specialization," as used here, refers to the users of the equipment. In non-specialized technology, practically anyone can use the equipment at the skill level required. Thus, according to Fernand Braudel, the chair (and furniture is plainly equipment) has an interesting history:

> The chair probably arrived in China in the second or third century AD, but took a very long time to become a general item of furniture . . . . It was probably European in origin, whatever detours it may have made to arrive in China (via Persia, India or northern China). Moreover, its original Chinese name, still current today, means 'barbarian bed.'
>
> It was probably first used as a seat of honour, either for lay or religious purposes. And even recently in China chairs were reserved for guests of honor and old people, while stools were used much more frequently, as in Europe in the middle ages.——But the important thing is the sitting-up position that the chair and stool imply, and therefore a way of life unlike that of ancient China and the other Asiatic countries, indeed unlike all non-European countries.[3]

This suggests that there is some skill involved even in using chairs, as the attempt to learn to live on the floor using cushions and carpets will confirm for chair users. And, of course, to drive an automobile, it is necessary to sit in the seat. The case of chair using suggests, furthermore, that a skill may be nonspecialized within a society but specific to that society in comparison with other societies, i.e., be culture specific. Then again, humans (and brutes) need at least a little learning in order to stay out of traffic by walking on the sidewalk and even simply to climb steps, but such skill is not specialized. The extent of non-specialized equipment use is probably underappreciated because of its ubiquity and because much of the equipment is passive and/or non-artifactual, but it is the technology in the background of all other technology.

(2) *Specialized Technology* may not be as old as non-specialized, which humans have had for millions of years. Specialized technology probably began with divisions of labor in foraging societies between adults and children and males and females, but, then again, perhaps that also goes back as far as *homo sapien sapiens* ought to be considered human, which is a matter of hundreds of thousands if not millions of years. Since then, many types of specialized *pre-machine technology* can be distinguished, especially in relation to agriculture and urbanization. Possibly it could all be called "craft technology." Here there are not only specialized skills, social roles, and groups, but also specialized equipment and localities for production and consumption. The equipment is relatively simple (although the flail, loom, and potter's wheel certainly have moving parts) and usually powered by humans. To be sure, brutes and also some inanimate power, e.g., the wind in sailing ships, was harnessed in ancient times, but the technological focus in craft technology is on the use of *hand tools* by humans. Even in America today, much of the domestic technology, e.g., cooking and eating with sinks, stoves, pots, pans, knives, forks, dishes, tables, etc., as nonspecialized equipment, is rarely beyond the craft stage, although of course much of the making or at least repair of clothes and the cleaning of clothes, carpets, dishes, etc. has recently been mechanized and even scientized through applied physics (microwaves) and applied chemistry (detergents). It deserves repetition that much more classification is possible with respect to what can be called *craft or specialized non-machine technology*. Like nonspecialized technology, craft technology here chiefly serves as a contrast with other sorts of technology. Nevertheless, there is a great deal of it in advanced societies still.

(3) *Machine Technology*. Driving a car is specialized socially inasmuch as younger children are excluded from it (not to speak of the difference between amateur and professional driving and special licensing for the latter, e.g., taxi and truck drivers), but it is more than merely craft technology when one considers the *mechanical complexity* and *non-human power sources* involved in the equipment. With such complex power technology there can be greatly concentrated and increased industrial production.

According to the work of Lynn White, Jr., equipment thus generally characterized as machines ("engines" could be an alternative expression but "machine" has the related modifier "mechanical") became technologically focal in the Western Middle Ages with the heavy plow, refinements in equipment for the use of horses, the water wheel, the windmill, improved ships, and the mechanical clock. (Once one recognizes that *machines* can include brutes as sources of power and, for that matter, guidance, as in the case of the horse that knows his own way home, one can begin to wonder whether humans can similarly be parts of equipment, e.g., the drivers as parts of conveyances such as taxies, trucks, and busses, but this question does not need to be pursued here.) The conquest of steam and then internal combustion are later substages of a stage of technological development that had dramatically accelerated practically a millennium before. Especially if electric power transmission is added, it may seem that most of the equipment humans in advanced societies live with today is of the machine type. But, then again, if one again ponders the extent of nonspecialized and craft technology and also non-artifactual and passive technology, this may not be so sure. With power machines or engines, society came to have mechanics and engineers in addition to craftspeople.

(4) *Scientific Technology*. The most recent and for many the most conspicuous sort of technology is or involves so-called applied science (so-called "high technology" is only a very recent type of scientific technology or "sci-tech"). For example, the scientific discovery of the effects of light on silver nitrate led to the invention of the camera and thereby the modern revolution in the use of pictures (and are not pictures equipment?). In this most recent subspecies of specialized technology, where there is greater skill, more extensive and intensive training processes, and ever more powerful and complex machines, the difference is in the incorporation of

scientific methods and results in the designing, making, exchanging, and perhaps even the consuming of the equipment. If "applied science" connotes the existence of the science before the technology, then it is a misnomer, for usually a long existent technology is improved through the *inclusion* of science. "Scientized technology" seems too ugly a name, but "scientific technology" is felicitous and readily subsumes scientific medicine, scientific bridge and road building, etc.

Only a few remarks about science are necessary here. Despite popular preoccupations with the immediately applicable, it might be preferable to support purely theoretical investigation because the *eventual* applications are greater. Second, it is unfortunate that "science" has come to denominate merely "natural" science, for certainly the human sciences, e.g., sociology and history, are sciences as well (and also "applicable"), in which case the noun "science" ought always to be qualified. Third, it is probably typical that scientific technologies incorporate methods and results from various sciences, so that there is no isomorphism between a taxonomy of theoretical natural and human sciences and a taxonomy of scientific technologies. Fourth, not all technological advance is in scientific technology today: the recent breeding of milk cows down to the size of large dogs to fit third-world situations involved techniques familiar for thousands of years. There is also serendipity, e.g., the case of the planting of the verges of superhighways with wildflowers, which was part of Lady Bird Johnson's highway beautification program, led to ground cover that required much less water and mowing and thus saves millions of dollars a year.

Nevertheless, how science fits into technology needs to be looked at more closely and that requires some analysis of equipment using.

2. The Constitution of Equipment

Technology qua use of equipment on the world can be analyzed phenomenologically. In such investigation, on the one hand, one reflectively observes, analyzes, and describes the intentive process of using as intentive to its immediate and mediate objects, i.e., one reflects noetically, and, on the other hand, one takes the same approach to the equipment and purposes as intended to, i.e., reflects noematically. Proceeding in this manner, the most conspicuous feature of reflectively observable equipment using is that it is predominantly *volitional.* Whether or not there was deliberation and decision behind a process of equipment using and whether or not it is actively engaged in or goes on habitually in what Husserl called secondary passivity, it is a striving or endeavoring. As such it can be positive or negative and, if firm, its resolute, or if shaky, its hesitant or tentative.

Correlative to the equipment using is the reflectively observable *equipment and the world it is used on as used*, which can similarly be operational or habitual, positive or negative, and resolute or hesitant. To be sure, only aspects of the world of a given society are thematized, but impacts can be rather broad, especially if the inadvertent ones are included. The subtle thing to grasp about objects as willed is their functional or *use character.* This can be intrinsic or extrinsic, i.e., it can be the end use or the means use. When the hose and soap are used as means to wash the car, the clean car is the end, but then again the car, clean or not, can be used to get to work or bring home food, in which case it has means use for the driver. The chains of ends and means can seem endless. It needs constantly to be asked just for whom and in relation to what is something a means or an end. And it needs to be recognized that the same item can for the same subject on different occasions or for different subjects on the same occasion have use characters of both the end character and means character sorts. Analysis becomes complex in other ways. For example, not only is the automobile active equipment by means of which one travels, for whatever purpose, but also the road is another and related means, another equipment, but passive. Finally, for this level, and returning to the noetic side, there is a difference to be recognized between the willing of ends and the willing of means to those ends.

While the volitional, just sketched, predominates in technology, there are also evaluational and cognitive components in the using of the equipment on the world. The general point may be easiest established with reference to the difficulties one encounters when attempting to will the creation or preservation of something one values negatively, e.g., buying a car one does not like. *Valuing*, like willing, can be positive or negative (set aside the neutrality of apathy), i.e., liking or disliking, and also firm or shaky. Remaining in noetic perspective, there is also a *cognitive component* in the intentive process. The cognitive is a matter of believing, which can be positive or negative (set aside the neutrality of skeptical doubt), i.e., believing and disbelieving, and also operational or habitual. The species of the genera of the firm and shaky in this "doxic" sphere are certainty and uncertainty.

Noematically, the object as evaluated, i.e., as liked or disliked, has positive or negative value characteristics, it is good or bad. Such values are intrinsic or extrinsic, i.e., X can be valued for the sake of Y and Y can be valued for its own sake. And analogously the object as believed has its *belief characters*, it is believed in or disbelieved in (in ordinary English, propositions are "believed" and objects "believed in"). Sometimes, one thing is believed in on the basis of belief in another thing, e.g., effects of causes and vice versa, and then there is intrinsic and extrinsic belief character to the objects. (If the end character of things willed for their own sakes and the means character of things willed for the sake of other things is remembered, it is seen that the intrinsic/extrinsic distinction holds in all three spheres.) In Constitutive Phenomenology, the belief character, when objectivated, is called "existence" (including "inexistence"), just as value characters in that event are called "goodness" and "badness" and use characters are called "usefulness" and "uselessness." Then one can go on to say that something not only exists or does not exist but also that it is good or bad and useful or useless.

Besides that which has just been sketched and which can also be called the praxic, pathic, and doxic modes of positionality, technology as using of equipment on the world includes *awareness* in the user. Correlatively, that which is used, valued, and believed in is also something of which one can be aware. For technology this is basically sensuous awareness, but when the equipment or what it is used on includes a psyche it is more than that. Automobiles are sensuous objects. One can expect to see a car, then see one, and later remember seeing it. The changes from "expected" to "seen" and from "seen" to "remembered" are changes in the car's manner of givenness. That which is thus variously given has color, shape, hardness, weight, smell, sound, etc. Plainly what a car is can be described in such terms, but insofar as it is equipment, i.e., fully what "car" signifies, the use character needs also to be described as intrinsic or extrinsic, positive or negative, firm or shaky, habitual or operational, etc. This is clear when equipment is reflected upon in correlation with the using of it.

An especially interesting aspect of equipment and other practical objects is their futurity. One can believe in and value objects in the now and in the past and, even if it is difficult for some philosophers to accept, one can believe in and value atemporal objects, such as numbers and propositions. One does not, however, will the atemporal (although one can will a future awareness of such an object), nor does one will either objects that are given in the now, i.e., as perceived, or objects in the past, i.e., as remembered. What willing is intentive to is something that is expected. This can amount to the continuance of something currently perceived, but then it is with respect to the same thing at some future time. And plainly it can readily involve some modification of that which is currently perceivable, including destruction. Then again, if the object is not now given, the willing can be to create it or to prevent it becoming thus presentable.

Often when something is said to be "expected" this signifies that it is believed possible and likely to be given in the now, whether or not it is already thus given either to the person "expecting" it or to another. Thus, if one saves money in order to purchase a new automobile in five years, the goal of the saving, i.e., ownership of a car not yet produced, is expected in this way, but the car to be owned is not believed actually to exist, much less to have previously existed. In contrast, one may own an automobile and *believe* it to be in its garage now as well as to have been there all night and in other places in the past. When one *uses* the car to go shopping, however, it is not its being given where it is or was but where it will be, i.e., at the store or back home with the purchases, that is willed. The valuing involved in using equipment is also *futural* for the user *while she is doing the using*. One prefers the car on the road going, at the store, on the road returning, and back in the garage later over the car continuing in the garage during the same future time. Then again, one can prefer that the car belong to someone else, and on that basis strive to sell it, or even prefer that it cease to be able to be given anywhere and on that basis strive to destroy it.

As suggested earlier, equipment as such is a means. Hence it can relate to other means and also to ends. Other means involved with using an automobile include roads, which are passive equipment in relation to it as active. But there is much more than that. There is an entire street and highway system, stoplights, traffic police and courts, technologies for designing, building, and maintaining such, etc., technologies for fueling the vehicles, etc., etc. Where ends are concerned, they might be classified as proximate, intermediate, and ultimate. One may use the car to get to work, but working is not an ultimate end if one works to buy food, which one does in order to eat, etc., and perhaps even surviving is not the ultimate end for an individual or a group. If the above description tends to emphasize individual equipment using, that emphasis can be adjusted by recognizing that there are many cases where not an *I* but rather a *we* uses equipment. Millions individually and in groups, e.g., covered wagon trains, use a national highway transportation system. The state police us it to make a living, as do truck dirvers. Some use it to get to a place for vacation fun.

Yet another dimension of analysis is valuable for understanding technology. This relies on the distinction between production and consumption traditional in economics. In a broad signification, the distribution of the equipment produced can be included in the production process. That would include not only sales rooms for new cars and used car lots, but also various advertising and financing techniques, as well as all that is involved in extracting and transforming the various materials into parts that are assembled into automobiles. The operation of designing cars can also be included in production in this broad signification. What then are the relations among the operations of designing, making, and selling cars? Plainly these are related such that the designing precedes and is for the sake of the making and the making is for the sake of selling cars. Thus there is an automotive industry.

An historical study of automobile technology would probably relate the automobile to earlier types of transportation equipment, such as the wagon, the stagecoach, and the carriage before it became horseless and focus on how the designing, making, and selling of such equipment became more and more complex, going from craftsmanship to a highly mechanized industry, and even now includes much science at every stage. But what might well be forgotten in such a treatment of the automobile is that *it is to be used by its consumer*. In the narrow signification, *equipment using* connotes this goal and is equivalent to consumption. It is nowise necessary that the consuming of the equipment destroy it. When the equipment is nonartifactual, e.g., the stones thrown at prey, there is consumption without production. The consumer or user may use the equipment for various purposes, perhaps as transportation, perhaps as an indicator of social status or as a means to an enhanced erotic life. There can be multiple purposes and if automobiles of various makes are effectively equal in one respect, perhaps buyers can be attracted through emphasis on other aspects. For simplicity's sake, it will here be assumed that the sole use is transportation, e.g., to get to work and to bring home groceries (but the analysis could be extended through recognizing primary and secondary purposes). If the car is for transportation, then that is what it is designed, built, and sold for. Yet given how little car using in the narrow signification is socially specialized and the passivity of so much of the rest of the transportation system, e.g., the sidewalks, highways, lines on the highways, etc., and how comparatively specialized, mechanized, and scientized the active designing, making, and selling of cars has become, the traditional focus of attention on production to the disregard of consumption can be understood.

Technology as equipment use can thus be analyzed with respect to the willing, valuing, believing and the awarenesses involved in the using and the use, value, and belief characters as well as the sensuous objects carrying them that make it up. Such an analysis can be conducted on every part of any simple or complex technology, e.g., the automobile production and transportation industries and systems, and all of the components seen to relate teleologically. Moreover, this analysis is general and thus can be specified for nonspecialized, craft, and machine as well as scientific technology. Technology can be analyzed further, e.g., resorting to what Husserl calls intentional analysis and genetic analysis. These few pages of sketchy analysis should be sufficient to clarify technology as equipment use and to make attempts to answer the questions of how scientific technology is scientific and of how scientific technology can be rational possible.[4]

3. Rational Sci-Tech?

Can technology be rational? This is a philosophical question that is advanced if it can merely be presented in a promising perspective. Some preliminary remarks are in order. First, it is best not to conceive of something as either rational or irrational. Instead, one can consider something as more or less rational in relation to other things. Second, things that may be rational in some degree are not intrinsically rational but are made rational by other things, which expressions based on the verb "to justify" may better connote. Then one needs to seek that which would justify or make something to be to some degree more rational than something else. Third, it is well to consider that sorts of technology other than scientific technology might indicate increased rationality. The wheel barrow, for example, requires one person and a wheel to transport the same load that previously two persons had carried in a box with handles at both ends. Except that the equipment would cost a bit more, it is twice as efficient in terms of labor, no adverse side effects are evident, and hence it is a more rational way to move loads, but nothing scientific is necessary to its design, construction, or use.

With regard to the automobile, there is plainly a great deal of science incorporated into the design, construction, and distribution of cars and science now can play a role in the construction of roads and bridges and other passive technology in the transportation systems of advanced societies and, as well, in the training and behavior of relevant personnel at all levels, including sometimes the drivers of cars. Yet the case of active car consumption, which, as argued, production serves, is not so clear. The skill involved is specialized and sometimes at a craft level, and a power machine is used. Consumption is focused on using the car for transportation, to carry people to work, to bring home food, etc. The consumption of equipment begins with acquisition of the equipment, with gaining use of it. Here the focus will be on buying a car as a mode of acquisition (one could also beg, borrow, or steal or even rent or lease). Not only is the operation of acquisition subordinate to using the car but other equipment is also involved in that act. The new car showroom or the used car lot are passive equipment and money or credit are active equipment. Probably it is wisest, however, to consider that the salesperson is more the automobile seller's than the automobile buyer's equipment.

It might be that buying a car can be done somewhat scientifically. If so, that might show how science is incorporated into technology. But first it can be asked whether non-scientific buying can be rational. Two common-sense maxims (certainly there are many more) by virtue of which the act in question is more rational are (1) Decide in advance what one wants! and (2) See for oneself!

In the first respect, by attempting to decide in advance, one resists being distracted by relatively irrelevant factors. It may be that one has loved blue things and hated green things since childhood. If one decides in advance however that a four door vehicle that can seat six adults is essential, then there is less chance of preferring a blue two-door over a green four-door.

To be somewhat less simplistic, one ought to get clear in advance on whether one wants a status symbol or transportation and then, if it is transportation, then just what sort of transportation is to be sought, e.g., around town or cross country.

In the second respect, one should experience for oneself the cars among which one is to choose. Direct experience of the thing is more reliable than hearsay from friends or representations in magazines or on television, which nevertheless do influence buyers and are perhaps not entirely useless for the process, at least insofar as they stimulate consideration of additional possibilities. Not only should one look at the vehicle but also get into it and drive it and indeed drive it under a relevant variety of conditions rather than merely around the block and also preferably without the salesperson there to distract one's attention from weaknesses that might appear. Finally, if it is a used car, one ought to have one's own auto mechanic examine it. Since she is working for the buyer rather than the seller, she can be considered an extender of the consumer's vision.

The foregoing may seem merely "common sense." Hopefully it is. The point is to suggest that, without being scientific, car buying as technology can be more rational in some respects than it would be otherwise. Closer analysis of seating capacity may make this clearer. Suppose the choice is between the two-door blue car and the four-door green car. It might look as if three adults could sit the front and in the back seats of both and the salesperson and the advertising might also assert it firmly. If, however, one found six adults and had them get into first one and then in the other of the two cars, one would more clearly see whether those assertions and impressions about capacity were true. Moreover, something would be seen about how easy the different vehicles were to get into and out of and how well they carried what might be more than the typical amount of weight.

With reference to the second part of this essay, the question of what reason is can now begin to be approached. Phenomenologically analyzed, the "seeing" practiced in the above case would function as *Evidenz* (since "evidence" in English regularly designates objects rather than the awareness in which they are given, "evidencing" is a better equivalent). *The epistemological role of evidencing is to justify believing.* Seeing for oneself how well the two cars carry six adults would justify believing one more capacious and/or stronger than the other when loaded that way. (How the car so loaded would handle at speed on curves in the rain would still be undecided, however.) Such a believing would be better justified in that case than the believing based on what the salesperson or the advertising said. Indeed, in ordinary English one would probably say that one *knew* which car could carry six adults more comfortably, safer, etc. The more adequate the evidencing the more rational the believing.

How science is incorporated into technology is now also clearer. Science differs from common sense in the sophistication with which it justifies believing beginning from evidencing. Through care in inference, it can justify believing with various degrees of probability where there is no direct evidencing. Suppose the six adults are often going to be carried at fairly high speed on torturous roads in bad weather. For their protection one is interested in such things as how crush resistant the body and how effective the safety belts would be if the car were to leave the highway and roll over a few times. It might be expensive and immoral, but beliefs in those respects could be decided by sending a few cars-full of people off the wet road at speed, i.e., by the same non-scientific type of experimentation that decided whether six would comfortable fit inside. Then again, there are science-like methods by which the forces acting on the relevant bodies can be estimated and beliefs then established without destroying any cars or killing any people. Technology can become scientific through the incorporation of scientific results, but it can also become scientific through using scientific, or more likely, science-like methods. Science-like methods are not strictly scientific methods because they are more particularistic and practical in their aims than is typical for science. Thus dummies might be fitted with sensors and cars driven off the wet curves by remote control. This would be more moral. It could be more than common-sensical seeing with respect to the control of conditions and perhaps even involve the testing of hypotheses. At the same time, it would hardly bring new theoretical understanding about inertia, gravity, centrifugal force, friction, or even the properties of rubber, nylon, glass, and steel. Much that is derived from science and applied to technology in the way of methods as well as results may not be very scientific. Probably it will become more so in the future. After all, scientific technology has only been seriously underway since about 1850.

If it is thus clear how the automotive engineer might be somewhat scientific in designing and testing cars, the question remains concerning the car buyer. The tendency to focus on production to the disregard of consumption needs to be resisted. Since few consumers can undertake science-like testing, the best available resort is to relevant experts. There are presumably independent government and private testing agencies whose results can be considered before a purchase is made. Automobiles are probably reviewed more carefully than films and the reviews often seem somewhat scientific when technical data are considered. And the lay consumer can cultivate suspicion about interested information, e.g., that which comes from the car producer and distributor or a government attempting to keep an industry afloat, and seek independent scientific information instead. This would again seem the most rational of the approaches available and perhaps it manages to earn the qualification "scientific."

In view of the enormous sophistication of science, one could be tempted to identify the rationality of scientific technology with the degree that it is scientific. This identification could be mistaken in two respects. In the first place, there can be conflicts between common-sensical and scientific "evidencing." Thus an engineer might develop a science-like account about tire traction on wet roads in various speed ranges and with a certain distribution of weight in the vehicle and on that basis deem the car safe, but an experienced driver can find that the car feels unstable under the specified conditions. Might it not be wiser to respect the latter evidencing at least to the extent of seeking more data? Nonscientific equipment using is not devoid of rationality.

In the second place, there is the question of *the relation between the cognitive and other sorts of rationality in technology*, which after all is a predominantly *volitional* process of equipment using. And even if the rationality of the volition is derived from the scientific or at least science-like believing in the foundations of the equipment using, the derivation can be examined. The mediator between epistemological and praxiological reason is axiological reason. When one subjects technology to examination or critique, one needs to examine evaluation as well as belief. One needs to consider values.

By the common-sense "experimental" method sketched above one can come to know that it is more difficult for six adults to get into and be carried by the blue two-door than the green four door car. By the principle of deciding in advance what one wants, a more effective conveyance is *preferred*. Preferring is a matter of valuing. Here it is also a matter of not intrinsic but rather *extrinsic* value, so that one of the two cars can be said to be better for carrying the six to wherever they need to be gotten. Their being thus conveyed is what has relatively *intrinsic* value, although it may also have extrinsic value, so that getting all six to the place together is good for something else. Evaluation is applied to what is known to be possible. In this perspective, one is examining values and valuing, not belief.

Valuing is often habitual, ill-founded, and thus prejudicial; thus the less rational alternative may get preferred. As mentioned, one might have a preference of blue things over green. If the utilitarian aspects of the car one would buy are preferred to the aesthetic, then that aesthetic prejudice would be diminished in influence. Valuing is more rational when one reflects in such a way as to recognize what the possibilities are and considers what leads to what, e.g., an overloaded and ill-balanced smaller vehicle to more accidents. Knowing that, one is right to disvalue it. Not being cognitive, valuing is not scientific intrinsically, but when based on scientifically established believing, it is scientific extrinsically. Just as evidencing justifies believing, making it knowing, knowing justifies valuing. Most of deliberation seems a matter of sorting out the values involved in the complexes of possibilities that are already known.

If willing is based upon valuing and the valuing is rational, then the willing is rational. Often the difference between valuing and willing is overlooked because action frequently follows automatically upon the settling of preferences. Yet one can prefer X to Y as conducive to Z and yet not go over to willing X as a means to Z as a purpose, which establishes that willing and valuing are separable and thus distinct. Just as the believing component within equipment using can be somewhat scientifically rational, so too can not only the valuing but also the willing that is founded upon it be rational, but in a derivative and then a doubly derivative way. Examination of volition not only involves consideration of the believing or knowing and the evaluation that would justify it, but also the review of the equipment and purposes to which one is already habituated. For example, writers familiar with one wordprocessing program tend irrationally to consider it normal and to consider faster and more powerful programs abnormal. But that can be corrected.

The examination or critique of technology begins in everyday life and can be carried to a philosophical level when one pursues ultimate justification for knowledge, evaluation, and willing. It centrally consists of an examination of the using of equipment on the world. Insofar as it weakens bad habits and strengthens good ones it has technological consequences. The uses of equipment may be scientific, or non-scientific, mechanical, crafty, and even nonspecialized. There is rationality on the level of volition and evaluation as well as cognition. The present discussion has been devoted to sketching a framework and approach to such matters. Further investigation will show whether other cases can fit this framework and whether sophisticated principles might be developed by which to judge technology rational or not. It is at least clear, however, that if technology is equipment use, then its critique is ultimately teleological. And for all the attractiveness of power machine and scientific technology in the production of equipment, production is for the sake of consumption.

Endnotes

[1]"The Historical Roots of our Ecological Crisis," *Science*, Vol. 155 (March 10, 1967), reprinted in *Machina Ex Deo* (Cambridge, MA: Massachusetts Institute of Technology Press, 1967), p. 77. According to his review of Needham in *Isis*, Vol. 75, White here depends on Daumas, "Rapports entre science et techniques: Etude generale du point de vue de l'histoire des sciences et des techniques, *Revue de synthese*, Vol. 83 (1963).

[2]*Technics and Civilization* (New York: Harcourt, Brace, & Co., 1934), p. 11.

[3]*Civilization and Capitalism 15th-18th Century*, Vol. I, "The Structures of Everyday Life: The Limits of the Possible," trans. Sian Reynolds (New York: Harper & Row, 1979), p. 289.

[4]This section has been developed phenomenologically from Dorion Cairns's unpublished phenomenological development of work by Husserl in value theory and ethics. Cf. Lester Embree, "Some Results of Cairns's Investigations into the Affective and Conative," *Research in Phenomenology*, Vol. 4, (1974).

TECHNOLOGY AND THE OBJECTIVITY OF VALUES*

Joseph C. Pitt

Abstract: It is suggested that a productive, i.e., useful, assessment of technology requires us to separate aesthetic value judgments (here used to encompass moral judgments, aesthetic judgments narrowly conceived, etc.) from cognitive value judgments. The rational determination of how best to handle the consequences of technological innovation requires that we determine the facts of the situation before we make policy judgments.

1. Introduction

Our task concerns the problems of learning to live with technology. Since the characterization of technology is itself an issue of some debate, let me begin with some preliminaries. I treat technology as humanity at work in the world. That is, technology is not a thing in itself, it is the techniques and methods, including machines, tools, social systems, etc., we use to make our way in the world. Given this perspective, let me now, for the purpose of focusing our efforts here, rephrase the objective of learning to live with technology in the following way. We are concerned with the problems created by the methods we use to manipulate and investigate the world. As such the philosophical problems of technology are problems associated with the reasoning we use to develop and use these methods

Lifeworld and Technology, eds., Timothy Casey and Lester Embree, Washington, D.C.: The Center for Advanced Research in Phenomenology and University Press of America, 1989.

and techniques and to assess the consequences, expected and otherwise, of these methods. These problems range over a variety of issues. I will be concerned only with the nature and role of values in assessing technology. And even then, my worries are narrow and restricted to the problem of structuring the debate over how best to assess technology.

Concentrating on how to assess technology raises the question of the appropriate framework within which to critically evaluate technological developments. As it turns out, there is more than one framework. Each is used under different circumstances, but they are all needed if we are to proceed rationally. The problem of evaluation is one of the major philosophical issues technology elicits and I am deeply concerned by the unproductive use of ethical frameworks in this domain. I will be arguing that the use of ethical categories to discuss technological developments is for the most part misapplied. This is not to say that under all circumstances ethical evaluations of technology are misplaced. Rather, I seek to differentiate ethical claims from other forms of assessment and to find an appropriate schema for indicating where each approach best functions. I am also not claiming that questions of values are not legitimate. But it is important to see that not all questions of value are ethical ones, and, further, that the rational assessment of technology requires an appeal to objective values. In what immediately follows I am going to construct an argument based on a discussion among a group of biologists on the issue of whether or not to continue recombinant DNA research. My reconstruction does not intend to faithfully capture the full flavor of the original argument; it attempts rather an abstract version of it for purposes of identifying a type of problem which is typical in debates about technology.

In a 1977 debate at the National Academy of Sciences (see *Research with Recombinant DNA*, National Academy of Sciences, Washington, 1977), Ethan Signer, Professor of Biology at MIT, produced an argument that amounted to the claim that using the techniques of recombinant DNA to develop more food is so risky it ought to be stopped. His argument went as follows: we are currently not doing all we already can to help starving people (perhaps implying by that we are not using the enormous grain surplus in America to its fullest). So, if we are not doing what we already can, why try something dangerous? Clearly the argument is confused. Signer runs several complaints together. First there is the worry over the failure to use our grain surplus to alleviate hunger. Then there is the second worry, that the use of recombinant DNA is dangerous. While the two worries are not necessarily related, they do share a common feature. They are both based on certain value judgments; first, about the appropriateness of using our grain surpluses in certain ways, and, second, about the risk involved in the use of recombinant DNA techniques.

Two points need to be emphasized. First, the claim that using the techniques of recombinant DNA is dangerous *is a value judgment*. When one hears it from a scientist one tends sometimes to forget that. But it is true nonetheless. Scientists make value judgments, even about the facticity of certain aspects of their own areas of research. This claim about the dangerous nature of recombinant DNA is not as outrageous as Robert Sinshimer's assertion three years earlier at the 1974 Assilimar conference. There Sinshimer argued that we should abandon any form of recombinant DNA activity because it was against nature. Nature, he alleged, has set a barrier beyond which we should not go. How he knew this was the barrier was not revealed, but that was the claim. Sinshimer has since modified his stance but it is important to note that both claims, that recombinant DNA is dangerous and that it is against nature, are value judgments made by scientists, judgments which exemplify the traditional is/ought fallacy. We will examine this further below.

But before outlining my second point let me add one further note here. It may be remarked that recombinant DNA research has proceeded at an incredibly fast rate and that given what we know today, it is unfair to criticize biologists who were worried ten years ago when virtually nothing was known about the potential results of such experimentation. In a sense that is true. My concern here is not with the rationality of the specific claims made in a state of relative ignorance but rather with the type of argument they represent: a form of intellectual bullying in which the authority of science is weaved into a value-laden position established independently of the facts of the case under discussion.

Returning to the Signer argument, the second point I want to emphasize is the inappropriateness of the ethical judgment about using U.S. grain reserves *as a criticism of recombinant DNA technology*. The fact that there are other ways to accomplish a goal does not in itself mitigate against one particular option. I will also return to this problem later. First, however, let us look further at values and value judgments.

2. Types of Judgments

Many types of value judgments have been made about the kind of threat posed by the new technology of recombinant DNA. These are primarily in the form of predictions about potential undesirable consequences if the results of recombination experiments do not pan out as we anticipate. These judgment/predications prey on our inability to know the full extent of the *risk* associated with results of this technology when it escapes human control. (See Sinshimer's comments in the 1977 Academy Forum, pp. 74-79.) The problem these judgments present is that they often obscure or confuse the facts of the case, thereby blocking our efforts to assess rationally the situation. This is not to say that value judgments play no role in assessing our technologies. The question before us concerns the kind of role.

Part of the problem here concerns our understanding of the nature of values and how they function in evaluative situations. New technologies, and some old, are often seen as presenting a threat to the life style (read "values") of some group or other. In learning how to deal with new technologies we need to learn how to assess the strength of such threats. But, more importantly, ultimately we are faced with the more difficult problem of determining the significance of the threat itself. Let me explain.

Objections to the introduction of new technologies are usually based on the claim that some valued state of affairs will be endangered. The snail darter was allegedly threatened by the construction of a dam. This dam was opposed because, largely ignoring the evolutionary history of species, that is, the fact that not all species survive over time even without man's intervention, some individuals made the value judgment that endangering the existence of a species is wrong. The fight that ensued was not over the efficacy of the dam, but over conflicting value systems with regard to the importance of a near extinct species versus the needs of the human community. In other words, the battle was over values. And, unlike battles over facts, there is no way to determine which of these values are correct.

If we had some account of values which showed how they could be as objective as facts, we might be in a position to short circuit many arguments over technologies. That is, if we could show that the preservation of species is an objectively good thing and not merely an opinion, we should be better off. But even that won't do. We need more.

We need both a theory of objectives values, that is, a theory which explains how values can be objective like facts, and, in addition, a theory which ranks values, i.e., which shows which ones are more important than others. If we had such a scheme, then if there were questions about whether or not we should proceed with some technological development, we could resolve them by appeal to the facts of the matter. The other option is to remove as much of the value dimension as possible from the discussion.

Unfortunately a theory of objective values is utopian. For one thing, it isn't obvious that values are there-in-the-world like events. If they were, surely by now we would have found them or have discovered some clue as to how to go about finding them. Second, there is some question as to whether or not *facts* are there in the world. On some accounts, what count as facts are a function of the theory being used to investigate the world. Hence, facts are descriptions of events or states of affairs, in the language of one theory or another. And, on an extreme account of this view, as theories change, so do the facts (see Kuhn, 1972). If facts change, then any search for objective values based on an analogy with the objectivity of facts breaks down. While I do not endorse the extreme version of this view, on which all facts change as our theories change, I do think that some sense can be made of the idea that some facts, those which are more theoretically based, change as theories do. (I will not develop this any further because it takes us afield into the murky lands of the realist/anti-realist debate which are not germane to our concerns here.)

Our other option for resolving conflict was to remove as much of the value dimension from the discussion as possible. Unlike our first option, which was utopian, this is just wrong-headed. To make progress we shouldn't ignore values–for that can't be done–rather, we must distinguish between those values about which rational discussion is unlikely and those about which we can proceed in a reasonably confident manner. The distinction I have in mind derives from Richard Rudner's observation that there is an entire range of values which we tend to overlook, but which still count as values (Rudner, 1965). These are *epistemic values*; the values which guide our search for knowledge. They include truth, objectivity, measurement, justification, etc. And they are to be distinguished from those values operative in the *other* area, the search for the good life, i.e., aesthetic values. I call these "aesthetic" values rather than ethical or moral values because I see ethical values as ultimately deriving from broader aesthetic considerations. It should also be noted in passing that it is not being claimed that these two categories, epistemic and aesthetic, exhaust the range of values. They are just the two under consideration here.

3. Epistemic Values as Objective Values

The difference between these two domains, the epistemic and the aesthetic, is striking. And while the values by which they are characterized function analogously to Kantian constitutive rules, in that they determine the legitimacy of actions, assertions, etc., within those domains, we are often slow to see that epistemic values are values nonetheless.

There are many explanations for this state of affairs. The fundamental one is that epistemic values are not as vulnerable to the open-question challenge as aesthetic values. Under scrutiny they may turn out to be as weak as their aesthetic counter-parts, but the initial reaction is not so. Epistemic values have a close connection with some sort of objectivity, meaning by that, with the way the world is. Thus, when we speak of a "true" statement, there is a sense that such statements relate to the world in a way which makes it possible to show without doubt that those statements have the property which they proclaim. The statement "snow is white" is true because the world really is the way it says it is. And it is that kind of "showability" that enforces the view that epistemic values like truth are immune to the open-question challenge. For what sense would it make to say, "Well, yes, the snow really is white, just like the sentence which you uttered and said was true claimed, but is that sentence really true?" This being the case, I want to argue that epistemic values are *as close as we can get* to objective values. That is, claims involving conflicting *epistemic* values can be resolved by a method which has a definite stopping point, namely the way the world is.

When we turn to particular cases, the situation is obviously more complicated than what has been suggested. But the line of attack here is straightforward. Epistemic values are *epistemic* values rather than aesthetic values precisely because of their relation to empirical knowledge. Remember that by aesthetic is meant not merely the values of art but also those of morality and ethics. Empirical knowledge is about the world, and if epistemic values fail to yield such knowledge, they cease to function in the epistemic domain. One way to see this is to look at the history of the theory of knowledge. The history of epistemology is the history of our efforts to discover those values which will most accurately guide our search for knowledge. And in our efforts to achieve a better characterization of knowledge over time, we have had to change values. Thus, starting in the seventeenth century we eventually gave up the search for knowledge as certainty because we discovered that, as a value, certainty was non-productive. There are other cases of value rejections and replacement as well. For example, one of Galileo's major contributions to the development of modern science was his insistence on the importance, i.e., value, of measurement as a way of distinguishing scientific truths from metaphysics. As he pointed out over and over again in the *Dialogue* and the *Discourse*, appeal to metaphysical principles must be replaced by rigorous measurement, for this is the only way to be sure of truth. In effect, Galileo was challenging the metaphysical principles used by 17th century Aristotelians with the open-question argument. Considering the following from the *Dialogue* where Galileo is attacking Aristotle's proof of the perfection of the world which appeals to the fact that the world has three dimensions:

> To tell the truth, I do not feel impelled by all these reasons to grant any more than this: that whatever has a beginning, middle, and end may and ought to be called perfect. I feel no compulsion to grant that the number three is a perfect number, nor that it has a faculty of conferring perfection upon its possessors. I do not even understand, let alone believe, that with respect to legs, for example, the number three is more perfect than four or two; neither do I conceive the number four to be any imperfection in the elements, nor that they would be more perfect if they were three. (Galileo 1632, p.11)

Galileo sought a method of providing reasons not open to the kind of objection he is raising here against Aristotle. And the general search for constitutive values in epistemology has generally proceeded in a similar manner, seeking ways to avoid charges of arbitrariness and metaphysical superficiality. The result has been a set of values that have come to occupy a place of honor, having survived the trials of experience (cf., N. Goodman, 1953, Ch.2). And to the extent that calling something "objective" is honorific, epistemic values are objective.

The point to be secured here is that truth is not just a property of a limited set of statements which meet certain logical conditions. By trial and error we have come to the conclusion that truth *ought* to be a property of those statements if they are to count toward knowledge. The same holds for other epistemic values. Thus, on normal accounts a statement does not have to be justified unless we want it to count toward knowledge. Likewise for the role of measurement, confirmation, evidence, etc. And the reason we have been able to select these values is that we know how to determine when we have knowledge, i.e., when these values have done their job.

Part of the reason we fail to recognize these concepts as values is that in epistemology we tend to use them in the formation of 'conditions of adequacy,' thereby disguising their normative features with an overlay of empiricism. But the reason they *are* conditions of adequacy is just because they meet our developed intuitions about what can and ought to count as empirical knowledge. This, however, is not enough to differentiate epistemic from aesthetic values. Aesthetic values also function as conditions of adequacy for exactly the same reasons. They too meet our developed intuitions concerning what can and ought to count as contributing to the good life. Nevertheless, there is a difference. For unlike the situation with aesthetic values, epistemic values do not lead us into an infinite regress, looking for the ultimate conditions of adequacy. For when all is said and done, epistemic values can be shown to lead to better or worse conditions of adequacy because what is supposed to count as knowledge can always be tested against the world. The test of knowledge is whether or not we can manipulate the world. There is no counterpart test for aesthetic values which escapes the open-question challenge.

Despite the efforts of some ethicists to consider the empirical consequences of ethical theories, moral categories are, for the most part, developed *a priori*. This is not necessarily bad. The articulation of moral categories and specific values is a response to typically abstract philosophical questions deriving from our normative search for an account of the Good Life. Those categories stipulate the context for moral and ethical discourse. That context is the world of the normative ideal, of what ought to be the case (cf. Rawls, 1971). Thus, ethical theories, unlike the theories of science, are not intended to be directly testable by daily experience. Furthermore, it is not clear what to do if when tested they are found to be flawed in some way. There seem to be two options: revise the theory or change the world. The problem is that there is no way to decide between the two and, to *prima facie* reason, to prefer one to the other. This general set of problems is exacerbated when we invoke moral considerations in dealing with the problems technologies create.

4. Problems With Ethics

Despite the fact that technology is the fruit of our labors, there is a major problem with using normative ethical categories to assess it. Approaches of that kind fail to address the technology in terms of whether or not it will do the job for which it was introduced. Instead, they confront the technology with a set of ethical standards, formulated under different constraints than were employed in its creation. Thus, rather than asking if the technology works as it was supposed to, we compare the effects of the technology to an ideal picture derived from some conception of what ought to be the case. In other words, we often tend to confuse two dimensions of our analysis, criticizing what is the case by reference to what ought to be and rejecting appeals to what ought to be by reference to the facts. The point here, however, is not that such evaluations should never occur; rather *it is not always clear that what it is claimed ought to be, really is what ought to be.*

Take, for example, the worry voiced by Signer over using recombinant DNA techniques to help grow more food. Signer's original argument was directed against the use of recombinant DNA because (1) no matter what science does to help produce more food, the problem of keeping people from starving remains a political one, and (2) we can't assess the risk of DNA, hence we should avoid incurring any risk. Here I would like to update Signer's worry and use his complaints to exemplify a general sort of problem. In particular, I will develop a Signeresque *scenario*—not be attributed to him—but compatible with the sort of approach he takes. To differentiate our scenario from Signer's original worry I will refer to it as S*. S*'s first objection is that we ought not to use recombinant DNA techniques because we are not doing what we already could be doing, i.e., using the enormous surplus of American grain to alleviate a problem, let us say the East African drought. Reconstructing the missing argument, let us assume the reasoning to be something like the following:

> (1) In those cases where the U.S. government feels it ought to alleviate some problem or other, we first ought to use available resources.
>
> (2) The U.S. ought to help.
>
> (3) The U.S. has available resources in the form of unused grain.
>
> Therefore (4) The U.S. ought to use its grain surplus to relieve.

There are several problems with this argument. The first is that it is not at all clear what decision procedure was used to pick the grain surplus as the appropriate available resource. Second, no thought has been give to the *consequences* of using our grain in Africa. This is notable since there was some consideration given to the consequences of using recombinant DNA techniques, namely, it was said to be dangerous. But there are also dangers associated with sending U.S. grain to Africa, three in fact. (1) Our grain may contain a biological contaminant which can be extremely destructive of native growth, such as the parasite which caused Dutch Elm Disease in the U.S. (2) There is the social and economic danger of creating a dependency on the U.S. (3) There is the social and economic danger of undermining the autonomy and stability of the way of life of the recipients of the aid. All three of these are real possibilities and yet they were not raised by S*. Rather, and instead, the specter of unloosing an alleged dangerous, man-made creation was invoked.

This argument against the use of this bio-technology is skewed for several reasons. First, it is incomplete. It fails to elicit the dangers associated with the choice of sending grain. Second, it asserts that DNA bio-technology is dangerous without considering the facts. Again, we recognize that in 1977 the facts were unclear. But I still am concerned with the type of argument S* exemplifies. It makes it *sound* like it is a fact that recombinant DNA technology is dangerous. Further, if we are to compare what we know about the probabilities associated with the occurrence of the *three* dangers elicited so far, we will find out that it is considerably more likely that we will create either a dependence on the U.S. or release a dangerous organism already in existence in the U.S. than we will create some kind of a monster agent using the technology of recombinant DNA. The reasons for this conclusion are several. First, the safeguards required then, as well as now, for the development and release of an artificial agent are not just strict, but considerably more strict than those for guarding against the other two dangers. Second, given the time it takes to develop and test an appropriate grain using recombination, the probability that we will have witnessed at least one of the three alterative scenarios will increase substantially.

I propose that S*'s solution to the problem of aiding the starving people of Africa, i.e., use our grain and avoid the dangers of recombinant DNA, typifies the kind of reaction one finds to new technological developments. S* did not obviously spend a lot of time thinking through the available options, examining them in the light of our knowledge about the kinds of problems they present and assessing their comparative merits. S*'s response mixed epistemic, aesthetic and even political-economic frameworks, resulting in a confusion of the facts of the situation and placing the discussion on the level of emotion rather than reason. In other words, we have here what is, by analogy with Gilbert Ryle's notion of a category mistake, a framework mistake. If the argument against DNA research is that it is dangerous, then this discussion ought to take place exclusively in the context of epistemic considerations, and then *all* possible dangerous outcomes ought to be considered–including the dangers of using U.S. wheat. If it is, rather, a question of moral obligation, then we ought to use an aesthetic framework–what should *not* be done is to start with one set of claims, i.e., epistemic, and then without warning shift to an aesthetic point.

The unintended shift to the aesthetic framework is signaled by suggesting as a solution to the problem that we ought to share what we have. It is then reinforced in the claim that recombinant DNA technology is dangerous. It isn't just that sharing is an ethical notion. Rather, in suggesting that by sharing our grain we will be doing all we can, we short-circuit the discussion of the available options and on suspicious grounds. This confusion of frameworks would remain even if S* had divided things up by suggesting a two stage process, one for the short term and one for the longer term, say, for example that what we can do in the short term is send grain, acknowledging that there is a certain danger in doing so and attempting to guard against it, while at the same time attempting to develop a long term solution to the problem. The confusion continues because we have eliminated the DNA option *a priori*.

By having a biologist describe recombinant DNA as dangerous we have created the illusion of a scientist asserting a fact, when he is merely asserting an emotional opinion. This is complicated by the fact that S* has behaved in a fairly unscientific manner by omitting the dangers associated with the option S* advocated. This provides the ammunition to defend my view that S*'s entire approach has been conducted in the context of an aesthetic framework, where advocacy does not necessarily reacquire full and considered examination of all the options. Rather, only when honesty is the supreme value will you be forced to give equal weight to all options. And in that case, there is no basis for choosing among them since honesty applies to the behavior of the investigators and does not reflect the merit of an option. Furthermore, in this case, S*'s choice to work within an aesthetic framework rather than an epistemic one, may have negative results because of the failure to deal rationally with the problem. Rather than proceed by adhering to epistemic values, the use of aesthetic values has confused and misdirected the argument.

5. The Methodology of Technological Decision-making

On the other hand, all is not lost. By distinguishing between epistemic and aesthetic values we have the basis for a methodology of technological decision-making. For now we can distinguish between two tasks: (a) discovering all there is to know about a given problem, including its history and the options open to us, with their relative merits and demerits, and (b) choosing among our options. In the process of discovering our options, epistemic values help constitute the framework for investigation. In choosing among our options, aesthetic considerations have initial priority. What should not be allowed is to have aesthetic considerations intrude into what is primarily an epistemic context. Thus, referring to recombinant DNA technology as "dangerous" is inappropriate in the context of attempting to determine whether that technology can provide an option and what its costs are. Everything we do is dangerous. What is important to know is not than an option is dangerous, but whether it has a greater potential for negative consequences than other options. However, what constitutes negative consequences is going to depend on the aesthetic values of the decision makers. Some consequences are more clearly negative than others, while other situations may produce results for which a choice is very difficult due to the comparatively equal weight of the values being threatened.

Does this mean that all technology related decisions are ultimately aesthetic ones; that it is impossible to be objective in our assessments? Not at all. If in the determination of our options and their strengths and weaknesses, we allow the conditions for knowledge to guide us and, if we follow those restraints as closely as possible, then the resulting situation will be one in which the nature of the conflict among our values, if there is such a conflict, should be clear. When we are clear about those conflicts, we are in a much better situation to decide rationally which course of action to pursue. Let me explain.

As noted earlier, disputes over what to do about technological innovations usually revolve around the perception that there is some threat by the technology to a life style or to a set of values. When the facts of the situation are as clearly before us as possible and our choice among our options clearly expressed, where there are differences they will not be over the facts. They can only be over the consequences of those facts for a certain privileged state of affairs. The desirability of that state of affairs is a matter of value. Only when those values are in the open can a rational decision be made. And at what level or in what framework is that decision to be made? I suggest a third framework–the rational.

There are then *three* stages to a good technological decision-making. The first state is *epistemic*, the second is *aesthetic* and the third is *rational.* Objectivity enters at two of those three stages: the epistemic and the rational. We have already seen how objectivity enters at the epistemic level in the form of epistemic values. I am now suggesting that objectivity also has a role in deciding between two options which differ because of their aesthetic components. In other words, *we can decide between aesthetic value judgments on rational grounds*. It is important that these value judgments are about predicted states of affairs. For this claim is not about decisions about value judgments about values.

When we speak of making rational decisions, we mean employing a principle of rationality to make judgments. The principle I favor is called the Common Sense Principle of Rationality (CPR). CPR is a model of simplicity; which is a virtue perhaps only in these contexts [see, Pitt 1988]. It only requires that we learn from experience. Thus when faced with deciding among options which differ by favoring one aesthetic value over another, and where there is no obviously superior option, CPR says consider what we have already learned, that is, consider other similar situations. Thus, we have learned in other cases that even with the best intentions in hand, we can disrupt an economy by undermining its cycle of production. Simply providing a starving nation with grain will not solve the problems that situation presents. We need to provide them with the means for feeding themselves. In the African case, we are not just faced with famine. The lack of rain has resulted in the consumption of the seed store. This may not seem too important, we can always send them seed. But there's the rub. American seed doesn't grow where it is sometimes needed. The destruction of the seed store means the termination of a hundred year old or more research project in applied selection by the African farmers. For decades they have been saving the best seed from each crop and using it to plant next years crop, saving from it the best in turn, etc. While American seeds have been selected for optional results in our environment, it doesn't follow that they will flourish in Africa. Finally, from past experience we know that recombinant DNA technology can produce good results, as in the development of artificial insulin and microbes that consume effluents. And as we carefully introduce artificially altered biological agents into the open environment, we have been increasing our understanding of the mechanics of recombination and the effect of natural selection on these agents. Thus, based on CPR, if we were limited to one option, we ought to choose to develop a grain using bio-technology which will meet African needs. But CPR also tells us that for a problem of this magnitude we have often needed more than one approach. So the result would be a decision to invoke a variety of solutions, some short term: feed them our grain now; others long term: send them our seed and see if it will work and also get to work on a research program aiming at an artificially altered seed which will survive in the African climate. Now granted, the solution proposed here has elements not available to Signer in 1977. But again, it is the kind of approach I am suggesting that has merit. Signer simply ruled out the recombinant DNA option because of unknown factors.

I have claimed that good technological decision making has at least three separate levels and that objectivity enters at two of them. At the first level, the epistemic level, objectivity is a function of the values we have developed which are constitutive of our concept of knowledge. Once we have uncovered all the options, we should make our judgments with regard to which ones fit our aesthetic considerations best. But those value judgments themselves need to be assessed and this is where CPR invokes our experience. We need to learn from our past decisions based on value judgments. For not to learn from our mistakes is to be irrational. It is in this manner that we hold out hope for the evolution of our aesthetic values away from *a priori* analyses toward the rigors of survival in an increasingly complex world. And this potential for rational assessment of aesthetic judgment is the new and compelling aspect of our technological age.

*My thanks to Richard Burian for his helpful suggestions on earlier drafts and special note of appreciation to Ronald Bruzina, despite my unwillingness to concede.

References

Galileo, G. (1632). *Dialogue on the Two Chief World Systems*, translated by Stillman Drake. Berkeley: University of California Press.

Goodman, Nelson (1953). *Fact, Fiction and Forecast*, Cambridge, Massachusetts: Harvard University Press.

Kuhn, Thomas (19OG72). *The Structure of Scientific Revolutions* Chicago: University of Chicago Press.

National Academy of Sciences (1977). *Research with Recombinant DNA*, Washington, D.C.

Pitt, Joseph. "Simplicity and the Aesthetics of Explanation" in N. Rescher, (ed.) *Aesthetic Factors in Natural Science* (forthcoming).

Rawls, J. (1971). *A Theory of Justice*. Cambridge, MA: Harvard University Press.

Rudner, R. (1953). "The Scientist *qua* Scientist Makes Value Judgments," *Philosophy of Science*, Vol. 20.

PATTERNS OF DISCOVERY OR SOCIAL CONSTRUCTION OF TECHNOLOGY: THE INVENTION OF THE TURBOJET

Edward W. Constant II

Abstract: In this paper, social constructivist and Kuhnian-derived community of practitioners based models for technological change are compared, using the turbojet revolution as an illustrative example. Each sort of model offers substantial insight into the processes of technological change; yet each does a relatively poor job of comprehending technological failures or the "paths not taken." This explanatory asymmetry may reflect an irreducible asymmetry in the nature of historical, that is, human, understanding.

The gist of this paper is to compare two, actually closely related, models or conceptions of invention and technological change, and to assess their strengths and weaknesses. All of the models I'm going to address, the two of central concern as well as their near kin, are themselves variants of a class of more general evolutionary or variation-retention models.[1] Being a historian, I will use a specific, albeit spectacular, episode of revolutionary technological change, the invention of the turbojet engine, to illustrate, and ultimately to evaluate, the two principal competing models for technology.

Lifeworld and Technology, eds. Timothy Casey and Lester Embree, Washington, D.C.: The Center for Advanced Research in Phenomenology and University Press of America, 1989.

This paper, then like Gaul, is really divided into three parts. First, I want to briefly describe "social construction of technology" models, with some reference to the tradition by which they were inspired, namely the self-proclaimed "empirical program of research" in the sociology of science. Second, I want to summarize in fairly schematic form what I know of the turbojet revolution–which is based on an earlier study I did–and in so doing, briefly present the model focused on the community of practitioners, that is, the essentially Kuhnian model for techological change, that I used there. And finally, I want to compare in some detail the social constructivist and the Kuhnian derived models for technological change.

Conclusions first, easily stated: while both the Kuhnian style "patterns of discovery" model and the social construction models do a good job of explaining success, the social construction models have the advantage of greater generality, especially the ability to better portray the involvement of multiple social groups in the definition of technology. Both types of models, however, share a common weakness of combining, in a rather *ad hoc* and unsystematic way, variation and retention (or selection) processes. Worse, neither approach does a very good job of explaining failure, a critical problem to which I will return at the end of this paper.

Because I have a historian's affection for chronological order, let me start with the "empirical program of research" or, in acronym, EPOR models from the sociology of science which originally inspired the social constructivist approach to technology. The essential point of EPOR is simply to explain adoption of scientific theories, or research programs or traditions, or consensus in science, without reference to the presumed veracity or verisimilitude of the knowledge so portrayed. Unilinear, progressive development is explicitly rejected. Science is a social process, no different, and no more epistemologically privileged, than any other social process. Scientific truth, then, is socially constructed by existential scientific actors: it exists solely as the result of their negotiation and agreement. EPOR's greatest success has come in microscopic, real-time portrayals of "laboratory life" and the way scientific problems, and results, are defined, constructed, negotiated, and "sold."[2] Beyond that domain, however, EPOR has had, in my opinion, serious difficulties. EPOR adherents squabble mightily among themselves over "reflexivity," or whether their own "science" of sociology is privileged.[3] Most lapse into ad hoc recognition of the epistemological efficacy of science, or a form of realism, and resist, although inconsistently, full-fledged conventionalism.[4]

There is perhaps less novelty here than meets the eye. For example, for a long time I think it has been recognized that Pascal's experiments on the weight of air and vacua did not logically refute Aristotelian notions of nature's *horror vacui.* A number of rhetorical and experimental moves were still open to adherents of the older theory. Rather, when Pascal marched up the Puy de Dome with a barometer, he conducted an *experimentum crucis* only in the sense that the relevant community had agreed in advance (that is, had socially constructed) the meaning to be attached to the outcome.[5] What seems to have compelled agreement on the interpretation of the experimental evidence, however, was not mere social convenience, but rather, in Mary Hesse's words, the "coherence of the whole in the theoretical network."[6] Now, I should perhaps pause here to note that these sorts of problems are not new to historians, although they have expressed them in different, and I think one has to admit, on the whole more graceful language. Over a half a century ago Herbert Butterfield published *The Whig Interpretation of History*, in which he lambasted historians' propensity to read teleology and unilinear "progress" into historical processes. The classic example is to see "religious freedom" and voluntarism as the triumph of democratic, libertarian ideals rather than as the wholly unsatisfactory outcome of sectarian bigotry and conflict.[7] E. H. Carr a generation ago adroitly addressed the problems of selectivity in history ("facts" versus "historical facts") and of the rather intractable circumstance that historians (and other social scientists) are inherently "participant observers." The problems EPOR seeks to address I think are deeper and more profound than just scientific positivism. To these general conditions I also want to return at the end of this paper.

The social construction of technology program, or SCOT in acronym, like EPOR in science, seeks to understand the invention, adoption, emergence, or definition of technological artifacts (or traditions of practice) without recourse to ostensibly "objective" properties such as efficiency, performance, practicality, or cost effectiveness. Emphasis is on "social construction" by relevant social groups who negotiate and define, quite literally, what a specific artifact is. For example, in a meticulously reasoned case study, Wiebe Bijker shows how the bicycle, far from just being "invented," meandered through a long series of changes and permutations during which diverse social groups, from manufacturers to boy racers to lady riders, "negotiated" what the artifact "bicycle" ought to be.[8] In an equally careful study of Bakelite, Bijker demonstrates the importance in the process of technological change of "technological frames" (outlook or perspective from which a problem is defined), degree of social "inclusion" (adherence to prior traditions of practice), and "interpretive flexibility" (ability to redefine what an artifact is).[9] In similar fashion, Boelie Elzen reveals the different meanings attached by different people at different times to the notion of an ultracentrifuge: for Svendberg, it was a highly specific solution to specific experimental problems in colloid chemistry; for Beams and Pickles in the United States, it was simply a very high-speed, general purpose rotary device.[10]

Other SCOT approaches are less highly structured. John Law and Michel Callon, for example, exploit the notion of "heterogeneous engineering" to describe the complex, multi-faceted nature of technological systems building. For them, as for Thomas P. Hughes (whose magnificent *Networks of Power* is less theoretical in bent, but more fully developed historically), technological systems must be understood as interconnected networks of natural, technological, and social elements, none of which merits primacy of place in analysis and none of which can be neglected. Thus, in a malevolently entropic world, system designers must continuously struggle against recalcitrant allies (electrons which short circuit, dynamos which break down, critical stock analysts, fickle consumers) and against overt opponents (gas companies), to maintain the integrity of their systems. Despite its sometimes fanciful verbiage, the great strength of the heterogeneous engineering approach is that it better captures the diverse elements intrinsic to large technological systems and the often extraordinary turn-abouts clever systems builders must execute. John Law, for instance, offers a striking portrayal of the way fifteenth and sixteenth century Portuguese navigators turned the Atlantic currents and winds from fierce opponents of navigation into treasured allies by inventing the "*volta*," or offshore circular route, to return to Iberia from the West African coast.[11]

On the whole, I find SCOT more persuasive than EPOR, from which it springs. I personally am not convinced by a portrayal of science which fails to account for, indeed refuses to acknowledge, science's obvious and special success in understanding and manipulating the natural world: it is, in Joseph Schumpeter's apt phrase, "like Hamlet without the Danish Prince." Put more concretely, and laying aside a lot of pseudo-open-minded, relativist hand-waving, I doubt that there are many people who, if their lives and that of their families depended on it, would trust Pat Robertson's predictions of a hurricane's track over those of the National Weather Service, however often the weather service has been wrong.

This is not to say that an account of science requires some realist or positivist foundation, just that any quest for, in Donald Campbell's term, the "social locus of scientific objectivity" must begin with a clear notion of what is to be explained, even if that explanation turns out to be the especially efficacious rituals associated with the practice of science.[12] I would also note here that I believe much of the efficacy of technology, at least since the early eighteenth century, derives from engineers and technologists having borrowed and exploited the methods of science, especially rigorous *ceteris paribus* testing and parameter variation, to conduct "vicarious explorations" in the development of technological artifacts.[13] That is, both science and technology depend methodologically on the ability to experimentally isolate systems, and to systematically vary or manipulate individual system elements or parameters (or theoretical variables) in order to discover causal relations. Such procedures permit technologists to exhaustively test new systems or new variants in controlled environments, rather than rely exclusively on crude trial and error: stress tests on a prototype airframe are cheaper, and more efficacious in advancing the state of the art, than plane crashes. SCOT is necessarily less assertive about the problematic nature of technological knowledge than is EPOR about scientific knowledge, for reasons laid out below. SCOT, I think, is better: it does much to capture the complex, heterogeneous, multi-dimensional character of technology.

Let me turn now to specific exemplars, inventors of the turbojet, which I hope will make all of this somewhat clearer. For the sake of brevity, rather than run through a typical historical narrative, I've grouped putative "inventors" of turbojets and its rivals into three sets: successful inventors, near misses, and wide-by-a-miles (Tables 1, 2, and 3, respectively).[14] Perhaps I should note here that these tables are not exhaustive of all relevant projects: not even all turbojet projects are included, nor are other reaction propulsion or turbine systems. Not discussed at all are the very large number of attempts to develop other alternatives to the petrol piston aeroengine, especially diesels, which together probably absorbed far more resources between 1920 and 1940 than did turbojet development. The descriptors of "near miss" and "wide-by-a-mile" are also a bit facetious. None of the projects in either category were, *ex ante*, intentionally directed toward creation of a turbojet as it ultimately emerged; all were trying to do something more or less different.[15] What all projects in each of the three sets depicted here do share is that all sought some radically different alternative propulsion system to replace the conventional piston engine and propeller. For brevity, the universe of all possible and proposed alternatives is here considerably condensed and simplified. The argument goes like this: Successful and, I should emphasize, multiple and completely independent (none of these men knew about each other) inventors of the turbojet (Table 1) shared the following characteristics. Socially, they were outsiders to the conventional aeroengine community, but well ensconced in the broader aeronautical community. All had training in, were practitioners of, or had direct access to advanced theoretical aerodynamics. While the community of aero-engine practitioners sought incremental improvement in the traditional system–the piston engine and propeller–and did so with extraordinary success from the 1920s right through the 1940s, successful inventors of the turbojet self-consciously sought a radically different solution. They did so purely on the basis of assumptions about possible performance derived from aerodynamic theory: they saw not failure in the normal technology, but the presumption (the presumptive anomaly) that it would fail.

Three core assumptions made the turbojet: first, that well-streamlined airframes could fly very near the speed of sound (over 500 mph, at a time when conventional aircraft rarely exceeded 300 mph); second, that, because of compressibility effects, the propeller could not function, or at least could not function efficiently, at such speeds; third, that, on the basis of airfoil theory, very high gas turbine component (compressor and turbine) efficiencies were possible. Thus a very light, efficient gas turbine used to produce thrust directly could propel an aircraft at very nearly the speed of sound. Significantly, successful inventors also made two other assumptions (which turned out to be correct) apparently unjustified by anything at all: that continuous combustion intensities theretofore never achieved could be sustained; and that high temperature materials would be available.

Let me emphasize that none of these originators of the turbojet had ever worked for or had any connection with a manufacturer of piston aircraft engines. No manufacturer of such engines independently or autonomously invented or developed a turbojet. Yet, quite ironically, none of the four original inventors of the turbojet managed to see his invention or original design through to production. Whittle's engine was taken over ultimately by Rolls Royce (a piston aeroengine firm) and by General Electric in the United States (who had manufactured piston engine turbosuperchargers). Neither the von Ohain nor the Wagner designs ever entered production. All production German turbojets were built by Junkers Engine or BMW, both of which had extensive experience in piston aeroengine production. What I think this pattern suggests is how the extraordinary systems complexity of modern production is nearly independent of what is produced—knowing how the production system works is more important, once the new type is understood, than invention itself. It is heterogeneous engineering, or social construction, with a vengeance.

A couple of things are noteworthy about the presumptive anomaly derived interpretation of the origins of the turbojet, which I at least am convinced is accurate in its essentials. First, like the Kuhnian model by which it is inspired, it is community of practitioners based: the key causal factors are adherence or non-adherence to normal practice and the perception on the part of radical inventors of presumptive anomaly derived from scientific theory. Second, it is asymmetrical: it does not (and possibly cannot) say anything about some unknown number of no doubt equally talented people who shared the same information field as the inventors of the turbojet, but did not themselves invent it. I think the SCOT model (especially Bijker's version) would do equally well in explaining the phenomenon, and possibly better because of its capacity systematically to incorporate extra-community groups (such as air ministries). Moreover, the SCOT or heterogeneous engineering approach ought to do much better in portraying the process of technological change downstream of actual invention, especially manufacturing and continuing development.

The second set, my near misses (Table 2) are a more heterogeneous lot. Two of the men are easily numbered among the most talented people in aeronautics: A. A. Griffith and Clarence "Kelly" Johnson. Griffith was principal scientific officer for the British Air Ministry and was responsible, together with Betz and Encke in Germany, for the application of aerofoil theory to axial compressors and turbines–the theoretical work which made any gas turbine system a realistic possibility. Johnson is perhaps the most brilliant aircraft designer the United States has ever produced. His designs range from the pre-World War II Lockheed P-38 Lightning to the current Lockheed SR-71 Blackbird reconnaissance aircraft. Each man came very close to independently inventing a turbojet, and the work each began led eventually to a turbojet project: the Metropolitan-Vickers F2 in Griffith's case, the Whittle-GE powered Lockheed P-80 in Johnson's. Yet each did not (I want to say failed to) invent the turbojet when it was plainly before him. I believe they failed simply because they were too closely wed to conventional notions of acceptable fuel economy as defined by the dominant normal technology. Or, in SCOT parlance, they suffered from too high a degree of inclusion.

Moss and General Electric is a stranger case. Younger engineers in the turbosupercharger[16] group wanted to try a pure gas turbine, but Moss, who had suffered acute disappointment in his own attempts to develop a gas turbine for power prior to World War I, adamantly discouraged them. Anxionnaz and Imbert, at Rateau, similarly came out of turbosupercharger development. Exactly what their performance assumptions were at this point is not clear; but France's collapse in 1940 doomed their efforts in any event. For these cases, the SCOT approach again may be marginally helpful. Moss' lack of interpretive flexibility and his life-long inclusion in the piston aeroengine community despite being deeply committed to gas turbines may account for his resistance to innovation. The Rateau case, in contrast, remains problematic: the "what ifs" dominate any judgment.

In general, however, what unequivocally demarcates success from near miss is grasp of the full set of assumptions described earlier as presumptive anomaly: clearly, mere brilliance and profound knowledge of the field, as evidenced by Griffith and Johnson, is not enough. Possibly it's just stochastic. Or, as I have argued elsewhere, more likely, recognition of presumptive anomaly is related, especially in the American and English cases, to differing national patterns in the pursuit and utilization of scientific knowledge: the Germans emphasized theoretical elegance and superior performance, while the Americans, and to a lesser extent, the English, pursued utilitarian, commercial development. These essential differences in the culture of aviation are embodied not only in the kinds of aircraft developed and in the kinds of aeronautical research conducted, but also in the mind-set of the relevant research institutions, and in the derivation of military air strategy and the specification of appropriate aircraft performance. Here something like the SCOT approach, generalized and lifted hierarchically several levels of analysis, from that of the artifact to that of broad cultural traditions, potentially could provide much better insight.

Finally, the wide-by-a-miles (table 3) are the theoretically most intractable group, no matter what theory is being applied. Guillaume actually patented what looks to be a turbojet. Yet its configuration, with multiple stage compressor and turbine, clearly places it in a much earlier steam turbine tradition, before application of aerofoil theory to turbine components; while the large starting crank protruding from the front of the engine in the patent drawings suggests that Guillaume was far from contemplating near sonic speeds. Moreover, Guillaume's patent appears at a time when inventive "projectors" (very much in Johnathan Swift's sense of the term)[17] were conducting what amounted to a morphological search (in Fritz Zwicky's sense)[18] for patentable alternative aircraft propulsion systems. In my opinion, to class Guillaume's invention as a turbojet, as sociologists ignorant of technology are wont to do, is simply to misapprehend what a turbojet is.[19] In contrast, Campini, in my view, was simply guilty of poor engineering. Whittle had "invented" a piston engine powered ducted fan system similar to Campini's in 1928, on his way to invention of the turbojet, and rejected it out of hand, and quite correctly, as too big, too heavy, and too complex.

Warren and Leduc are at the opposite extreme, seemingly reaching too far. Warren had worked on supersonic flow in steam turbine nozzles at General Electric during the 1920s (in a separate division from Sanford Moss), and reasoned that the shock waves which formed under such circumstances could be used for compression in a supersonic ramjet. Although his employers gave him considerable time and resources to work out his ideas, ultimately he and they decided that GE simply didn't have the market entree or technological expertise to enter such an advanced and problematic field. What is so ironic about Warren is that he was an expert in turbine systems, and with any exposure to aerofoil theory should have been but a Eureka away from the turbojet.

Leduc also proposed a supersonic ramjet as early as 1933, and exhibited a model in 1938. A radically modified version of his design was actually built post-war, but remained sub-sonic and proved inefficient and unsuccessful compared to by-then available turbojets. The SR-71 is in fact a turbo-ramjet, but it is a vastly different thing from what Leduc proposed. In retrospect, neither Warren nor Leduc had any real idea of the problems inhibiting supersonic flight,[20] and it is unlikely that their systems would have ever proved successful: in many ways, their ideas are like Guillaume's (and like many of the "inventions" attributed to Leonardo da Vinci), merely linguistic or artistic time travelers from another technological epoch.

What has SCOT to say of such failures? To suggest that Leduc simply failed to successfully "enroll" supersonic airflow as an ally of his system, or that Campini's rhetoric failed to convince the appropriate social groups (in fact, it garnered him considerable support), or that he suffered too much interpretive flexibility, rather than that his engineering was just objectively inadequate, is to me not very enlightening. Neither can I readily find demarcation between success and failure in technological frames or in relative degrees of inclusion: *ex ante*, those who would be successful and those who would not were equally *outré*. Bluntly, it does not appear possible to account for any of the misses-by-a-mile without reference to objective reality–as of course we recognize it currently.

A couple of moves are possible. One could invoke some notion of a time and path dependent[21] technological ensemble–an updated notion of "state of the art," or, phrased negatively, "the time not being right." For technology, such a notion would be analogous to Mary Hesse's (also time dependent) holistic theoretical network, founded ultimately in the evolutionary validity of our cognitive apparatus, against which or within which specific theoretical innovations are tested. Thus, in that time and path dependent world, the misses-by-a-mile simply were objectively impossible or inadequate.

A problem remains, however. Remember that the assumptions which lay behind the turbojet revolution were well-founded in presumptive anomaly derived from theoretical aerodynamics, except for high temperature materials and combustion intensities. It is impossible *ex ante* to demarcate between these unfounded assumptions which turned out to be valid and the equally unfounded assumptions underlying unsuccessful systems. Indeed, it is not easy to account for why the former were valid *ex post facto*.

Alternatively, one could resort to the brute fact that, unlike scientific theories, technological artifacts explore the environment directly, not vicariously. That is, technology is subject to direct environmental elimination: bridges fall down, engines explode, airplanes crash. Indeed, the first serious attempt to break the "sound barrier" killed Geoffrey de Havilland, Jr. (son of the aircraft firm's founder), and even the irrepressible Chuck Yeager very nearly died in his own successful attempt. When technological systems do fail, what went wrong is not always obvious, but that something did is usually pretty unambiguous. Indeed, learning by failure, or by trial and error, is perhaps the central element in the epistemology of engineering, as it is in other learning processes.[22] Still, disaster is a crude metric for demarcating success from failure: it cannot alone distinguish the better of two less than catastrophic systems, a fact which, without access to other means of testing, would essentially eliminate one of the basic means to technological progress. Moreover, "failure" in very complex macro-systems is not easily identified, nor are all potentially catastrophic remote consequences, such as suspected fluorocarbon depletion of atmospheric ozone, even conceivable or perceptible without recourse to esoteric science.

I think the problem comes down to this: Because we are, in Donald Campbell's phrase, "epistemologically deprived," the product of time and path dependent evolution (biological and social), there are irreducible asymmetries in what we can know. These asymmetries take several forms. Other than by simply positing stochastic processes of association, we cannot account for the statistically large number of people, apparently exposed to the same information fields as the protagonists of the turbojet revolution, who simply, or perhaps in fairly sophisticated and novel ways (the historical record cannot tell us), missed it. We cannot argue with any conviction, or without reference to *ex post facto* outcomes, that the unfounded assumptions made by successful inventors were any more sensible than those made by their unsuccessful rivals. Most critically, we are not very good at the conjecture of entire counter-factual universes in which to play out the complex consequences of alternative paths: how would Campini or Griffith have fared absent the turbojet revolution? We don't know, and to the extent that we can imagine, we don't know the prospective validity of our imaginings. At best, we know only what we and our ancestors have learnt and encrypted in our genes and in our culture, meandering through past environments. Certainly we are often capable of marvelously reconfiguring that knowledge to adapt to, indeed create, new environments–but because it is a time and path dependent process, we cannot know in advance for sure when we can and when we can't.[23]

None of this weeping and wailing is meant to retreat into the poverty of historicism, or into the world view Voltaire so nicely disposed of in *Candide*. Rather it implies, first, a certain epistemological modesty in what we assert we can or do understand; and, second, by extension, a certain tolerance of missteps, sociological and technological. Indeed I think the social construction of technology program has made an extraordinarily useful contribution to understanding technology by forcing attention to social elements in its development. I am not convinced, however, that SCOT can either dispense with reference to an external physical, and technological, reality or that it can account wholly for the path of technological development.

I'm not sure that the next move for the social constructivists ought not to be away from the micro-level social construction of "facts and artifacts" toward a hierarchically higher level social construction of processes of selection–the variation-retention rituals of science and technology, and their common culture. Clearly social construction of individual facts and artifacts occurs within a highly evolved and highly structured context of material and ideological culture. There is almost certainly an ongoing, virtually dialectical interaction between specific episodes of technological (or scientific) change at the microscopic level of social analysis, and the content and meaning of the broad, overarching culture (more than the sum of its parts) of science and technology. The next task, I would think, for the sociology of technology is to understand how that hierarchically higher level of the culture of technology is itself constructed out of the lower level processes of the social construction of artifacts, and how that culture in turn, and over time, constrains and channels the lower level processes. Such a broader, multi-level view of social construction would permit of locally stable equilibria (that is, relatively long term, unproblematic traditions of technological practice) while still accounting for the meandering, contingent quality of technological change.

Finally, in conclusion, let me suggest one additional step. I suspect that if one were to define technological systems very, very rigorously, simulations of successive generations of such systems could be run sequentially against one another.[24] In such a manner, it might be possible both to partially capture counter-factual universes, as well as to sort out the "objective" elements in technological selection from those that are social or conventional.

PROJECT	F. Whittle	H. von Ohain	H. Wagner	H. Schelp
ORGANIZATION	Power Jets, Ltd.	Heinkel Aircraft	Junkers Aircraft	RLM (German Air Ministry)
DATES (INV/DEV)	1929/1936	1934/1936	1934/1936	1938
SCIENTIFIC ASSUMPTIONS	aircraft speed, altitude/compressor, turbine efficiency	same	same	same
PERFORMANCE TARGET	near-sonic, high-altitude flight/ light weight	same	same: broad band analysis of turbine systems	same: broad band analysis of reaction propulsion
DESIGN	centrigugal compressor, axial turbine	centrifugal compressor and turbine (axial later)	axial compressor and turbine	axial compressor and turbine
PRODUCTION	(Rover), Rolls-Royce, GE (US)	Prototype; more advanced engines incomplete at end of war	to Heinkel; original design abandoned	Junkers Engine (004) BMW (003)

Table 1: Turboject successes

PROJECT	A. A. Griffith	C. Johnson	S. Moss	R. Anxionnaz R. Imbert
ORGANIZATION	RAE	Lockheed Aircraft	GE Turbo Division	Societe Rateau
DATES (INV/DEV)	1929/1936	1940	1931/1939	1939
SCIENTIFIC ASSUMPTIONS	aerofoil theory applied to compressor and turbine	near-sonic speed at high altitude	compressor and turbine efficiency extrapolated from turbo-chargers	high-speed?
PERFORMANCE TARGET	power and fuel efficiency comparable to piston engine; much lighter, smaller	high speed; competitive fuel economy	competitive fuel economy at high speed	?
DESIGN	contra-rotating, contra-flow axial turbo-prop	axial/reciprocating compressor; axial turbine, jet	centrifugal compressor/ centrifugal or axial turbine, jet	axial compressor and turbine, jet
PRODUCTION	too complex, failure; work basis for Metropolitan-Vickers F2 turbojet	too complex; development terminated at end of war	overtaken by Whittle design, 1941	development delayed by fall of France, 1940; not completed until after the war

Table 2: Near misses

PROJECT	G. Warren	S. Campini	R. Leduc	C. Guillaume
ORGANIZATION	GE Steam Turbine Division	Caprioni Aircraft	--------	?
DATES (INV/DEV)	1929	1930/1940	1933/1938	1921
SCIENTIFIC ASSUMPTIONS	supersonic flow in steam turbine nozzles	propeller inefficiency at high speed	supersonic flight	?
PERFORMANCE TARGET	supersonic: 1350 m.p.h. @ 80,000 ft.	high speed flight	supersonic flight	?
DESIGN	ram-jet	piston engine powered ducted fan with afterburning	ram-jet integral to aircraft	multi-stage axial compressor/ multi-stage axial turbine
PRODUCTION	proposal only; Warren diverted to work on steam turbines	system too complex, large, heavy, and inefficient	original design never built; much modified (subsonic) post-war design failure	never built patent only

Table 3: Wide-by-a-mile

Endnotes

[1]The best description I know of such models is Donald T. Campbell, "Evolutionary Epistemology," in P. A. Schlipp, ed., *The Philosophy of Karl Popper* (La Salle, IL: Open Court, The Library of Living Philosophers, 1974) Vol. 14-I, pp. 413-463. See also Donald T. Campbell and Bonnie T. Paller, "Extending Evolutionary Epistemology to 'Justifying' Scientific Beliefs (A Sociological Rapprochement with a Fallibilist Perceptual Foundationalism?)," in E. Halweg and C.A. Hookers, eds. *Issues in Evolutionary Epistemology*, The Newcastle, N.S.W. Conference of July, 1987 (Albany, NY: State University of New York Press, 1989); and David L. Hull, "A Mechanism and Its Metaphysics: An Evolutionary Account of the Social and Conceptual Development of Science," *Biology and Philosophy,* 3 (April, 1988): 123-155.

[2]Trevor Pinch, "Some Possible Implications of the Sociology of Scientific Knowledge for Our Understanding of New Technology: The Case for a New Sociology of New Technology," Paper presented to the Industry and Employment Committee of the ESRC, November, 1985, gives an excellent overview of the schools within EPOR and their relation to SCOT; see also Trevor J. Pinch and Wiebe E. Bijker, "The Social Construction of Facts and Artifacts: Or How the Sociology of Science and the Sociology of Technology Might Benefit Each Other," in Wiebe E. Bijker, Thomas P. Hughes, and Trevor J. Pinch, eds., *The Social Construction of Technological Systems: New Directions in the Sociology and History of Technology* (Cambridge, MA: MIT Press, 1987), pp. 17-50; S. Woolgar, "Laboratory Studies: A comment on the state of the art," *Social Studies of Science* 16 (1982), pp. 481-498; Michel Callon and John Law, "On Interests and Their Transformation: Enrolment and Counter-Enrolment," *Social Studies of Science* 12 (1982), pp. 615-625; John Law and R. J. Williams, "Putting Facts Together: A Study of Scientific Persuasion," *Social Studies of Science* 12 (1982), pp. 535-558; for a physicist's astute and angry critique of the EPOR approach, see Laurie M. Brown, "High Energy Physics: Claims and Constraints," review of *Constructing Quarks: A Sociological Study of Particle Physics* by Andrew Pickering, *Science* 228 (17

May 1985), pp. 857-858.

[3]H. M. Collins, "Stages in the Empirical Programme of Relativism," *Social Studies of Science*, 11 (1981), pp. 3-10.

[4]My (admittedly dim) perception of these issues comes from Mary Hesse, "Changing Concepts and Stable Order: review of H. M. Collins, *Changing Order: Replication and Induction in Scientific Practice*, in *Social Studies of Science*, 16 (November, 1986), pp. 714-726.

[5]See E. J. Dijksterhuis, *The Mechanization of the World Picture* (London: Oxford University Press, 1961, original Dutch edition, 1950), pp. 444-457.

[6]Hesse, note 4, p. 720.

[7]Herbert Butterfield, *The Whig Interpretation of History* (New York: W. W. Norton, 1965, original edition, 1931), especially pp. 75-89.

[8]See Pinch and Bijker, note 2.

[9]Wiebe E. Bijker, "The Social Construction of Bakelite: Toward a Theory of Invention," in Bijker, Hughes, and Pinch, note 2.

[10]Boelie Elzen, "Two Ultracentrifuges: A Comparative Study of the Social Construction of Artefacts," *Social Studies of Science* 16 (November, 1986), pp. 621-662.

[11]John Law, "On the Social Explanation of Technical Change: The Case of the Portuguese Maritime Expansion," *Technology and Culture*, 28 (April, 1987), pp. 227-252.

[12]Donald T. Campbell, "Objectivity and the Social Locus of Scientific Knowledge," Presidential Address to the Division of Social and Personality Psychology of the American Psychological Association, 1969.

[13]See Walter G. Vincenti, "The Air Propeller Tests of W. F. Durand and E. P. Lesley: A Case Study in Technological Methodology," *Technology*

and Culture 20 (October, 1979), pp. 712-51, on parameter variation as a central tool of technological development.

[14]This entire section is based on my *The Origins of the Turbojet Revolution* (Baltimore, MD: Johns Hopkins University Press, 1980).

[15]I am indebted to a participant in the conference, whom I could not identify at the time, for correcting my rather too facile summary of other projects.

[16]A turbosupercharger, or turbocharger, is a gas turbine driven rotary compressor used to supercharge (cram more air into) a conventional piston (petrol) or diesel engine.

[17]See Jonathan Swift, *Gulliver's Travels.*

[18]Fritz Zwicky, *Morphological Astronomy* (Berlin: Springer, 1957); and, "Tasks We Face," *Journal of the American Rocket Society* 84 (1951), pp. 3-20.

[19]W. F. Ogburn and D. Thomas, "Are Inventions Inevitable?" *Political Science Quarterly* 37 (1922), pp. 83-98; most of the alleged "multiple inventions" listed there are not identical; see Constant, "On the Diversity and Coevolution of Technological Multiples: Steam Turbines and Pelton Water Wheels," *Social Studies of Science* 8 (1978), pp. 183-210.

[20]Richard P. Hallion, *Supersonic Flight: The Story of the Bell X-1 and the Douglas D-558* (New York: MacMillan, 1972).

[21]Paul David, "Understanding the Economics of Qwerty, or Is History Necessary?" (American Economic Association Meeting, Dallas, Texas: December, 1984), provides superb insight into the time and path dependency of technological change.

[22]See David P. Billington and Robert Mark, "The Cathedral and the Bridge: Structure and Symbol," *Technology and Culture* 25 (January, 1984), pp. 37-52, for a critical discussion of the role of failure in engineering

learning. See also Carl W. Condit, "Another View of 'The Cathedral and the Bridge,'" *Technology and Culture* 25 (July, 1984), pp. 589-594, for an angry rebuttal of some of Billington's and Mark's more ambitious symbolic claims.

[23]Those familiar with his work will recognize more than a little of Donald Campbell in these sentiments.

[24]This idea is suggested in very general terms by William Todd, *History as Applied Science: a Philosophical Study* (Detroit, MI: Wayne State University Press, 1972), something akin to objective comparison of predecessor and successor systems was attempted by *Project Hindsight*, Office of the Director of Defense Research and Engineering, Department of Defense, 1969.

This is a very weak essay superficial.

ARCHITECTURA ARCHITECTURANS: WORLD(S) IN THE MAKING

Ronald Bruzina

Abstract: Clearly not all of the elements that are at play in the work of the architect are subject to his/her agency. One of the most important is the way the architecutral whole is a place and a whole of places—a kind of small world within the larger all-encompassing world. This essay explores the meaning of "world" in the basic ways it may be structured—i.e., by the fixing of places either statically or dynamically—and applies that analysis of architecture, so as to make clear how a/the world is present in its action of making.

There is a lot about architecture that is never studied either by architects or by those who, while not themselves architects, make a specialty of analyzing buildings and their history. Architects work, and architecture is done, always in particular places and at particular times and in particular societies—that is obvious; and this whole phenomenon forms the subject matter of architecture history and theory. But even in these studies some things remain implicit rather than explicitly focused on and analyzed. My aim in this paper is to draw out one of these implicit elements present in the work of architecture and intrinsic to the basic importance architecture has for human lives, but which is rarely directly considered. In the course of doing this I shall gradually explain the ambiguous and esoteric-sounding title: Architectura Architecturans: World(s) in the Making.

Lifeworld and Technology, eds. Timothy Casey and Lester Embree, Washington, D.C.: The Center for Advanced Research in Phenomenology and University Press of America, 1989.

1. The Work of Architecture as the Work of Origination.

According to centuries old usage, the *arkhi-tekton*—the chief builder—the architect, is the master in charge of the planning and construction of the building.[1] In our own day, the architect is indeed in charge of the planning of the building, and is the originator of its governing conception; but unlike in an earlier age it may not be the architect who actually oversees the construction of the building and ensures its execution according to that originating conception. That work of actual construction belongs rather to the engineering or construction firm. The architect today is principally responsible for the *conception* of the building, the *design* of the building. However minute the detail of that design may be, however far it may go in specifying, say, materials or principles of construction or even furnishings, it is the *creative conception* of that design that is the primary responsibility of the architect. The architect creates the originating, integrating conception that determines all the rest, as far as any such determination may carry. The architect is a kind of god with respect to that which gives the building its unity, its identity, its form, in the broader sense of the word. It is out of this "god's" creative soul that the idea of the building springs, whether slowly through stages, or in a sudden flash of imaginative genius. The actual construction of the building, then, transforms this idea into reality.

Having said all that, we immediately have to add, of course, that it is far from the whole story. The architect may be like a god in this act of creatively originating the conception that governs the realization of the building, but in all other ways the architect is most *unlike* a god. One need only mention that no self-respecting *god* would ever have anything to do with such a thing as a *client*. More seriously, a genuine god need only *think* a thing, and it *is*—whereas no architect has yet attained that kind of sheer professional power. There are, however, more subtle ways in which the architect simply is not in command, not even of the creative process itself which is supposed to be the core act of the architect's own identity as architect. For the power of imagination that is active in the architect as individual is not absolute. It comes to actuality necessarily within a context of lines of custom, interest, and purpose. It draws its vocabulary of formative elements from a concrete tradition, however much it may refashion, recombine, and challenge them. Finally, in creating a new architectural whole, a new building, the architect's imagination must have a community for which that new architectural reality can after all be meaningful and functional.

The history of architecture demonstrates over and over again the multiplicity and variety, and indeed the ultimate transitoriness, of the architectural "styles" to be seen typifying, in creative diversity, the individual works of a period or a geographical area. However difficult it may be to draw lines of absolute division, there is no mistaking the fact that, for example, the architecture of 11th century Romanesque churches displays a different "family" of forms from those of 13th century "Gothic," or those of the 15th century Renaissance—even in the same country, the same region, the same city—not to mention how different 11th century churches in France or Italy are from, say, mosques in North Africa, or temples in Cambodia. Whoever the architects were who planned these structures, they were guided in the definition and detailing of the respective governing conceptions by a trans-individual, culturally specific "matrix" that conditioned the forms of and gave cultural identity to the resulting new architectural masterpieces.

In sum, we realize that for architecture—as surely for all the arts—there is both a) the factor of individual creative capability, by which the specific structure is decided in individual disposition and detail, and b) the context of general patterning, by which the relevance and functional intelligibility of the resulting building can be grasped in the community. But once we understand this, we are then forced to raise the question of just what it is that determines the *power* and *worth* of architecture: is it the individual architect's creativity, or the cultural matrix? And before even attempting to answer that question, we would first have to settle the issue of how to understand the same concrete structure as, on the one hand, "functional" and, on the other, "aesthetic." What is architecture, after all?

This body of issues, and the considerations that lead up to it, are what is suggested by the expression "Architectura Architecturans." A literal, if awkward, translation of that expression would be "architecture architecturing," but that is little help. What is named in this expression, contrived after the model of a term of long philosophic lineage, is that total power which is the essence of architecture in its boundless creative dynamism. But, having said that, the question still remains: what *is* this "essence of architectural dynamism" that is ultimately at work in architecture and the source of its meaningfulness and validity?

The expression "architectura architecturans" is modeled after a famous term brought to prominent usage by Spinoza in the 17th century. Spinoza had an extraordinary conception of reality as all forming one single totality, whose internal differentiation was integrated in unbroken unity by absolute logic-ruled derivation. The supreme being and the whole of the universe were but a single substance, which nonetheless manifested itself in two fundamental modalities of articulation: matter in manifold extension, and thought in the concatenation of ideas. To each differentiated particular of the substance of this supreme being in one modality—in material extension—there corresponded exactly a differentiated particular in the other modality—in the concatenation of ideas. The term Spinoza used, now, to name this absolute substance in its totally integrated being, in its absolute self-sufficient power, was "Natura naturans"—nature naturing, nature as the being of self-unfolding power, of self-developing and self-articulating power. Just to complete the picture of Spinoza's conception which, though magnificent, is perhaps too cosmic and absolutist for most thinkers—we should add that the result of the self-unfolding, the self-development and self-articulation of this absolute being, the result of the working of this absolute dynamism, is the totality of differentiated particulars in extension and thought: the universe of material things and the universe of ideas, the world of sensuous space and the world of the multitudinous mind. This totality of differentiated things and thoughts Spinoza named "Natura naturata"—nature natured, nature as product of the working of absolute self-unfolding substance.

> [B]y "Natura naturans' we must understand that which is in itself and is conceived through itself; that is, the attributes of substance that express eternal and infinite essence; or God in so far as he is considered a free cause. By "Natura naturata" I understand all that follows from the necessity of God's nature, that is, from the necessity of each one of God's attributes; or all the modes of God's attributes in so far as they are considered as things which are in God and can neither be nor be conceived without God.[2]

If, now, we make the transposition to architecture, and in place of "the absolute substance," which in Spinoza's conception is God, put the total reality of the architectural work in its source and in its power–whatever in the end we would conclude that is–then we have a term like "architectura architecturans," i.e., architecture architecturing, architecture in the being of its self-articulating power. And we can see at once that the architect is only a part of this whole, an important and indispensable part, but nonetheless an agent who draws efficacy and direction from sources of meaning and figuration from beyond his or her own individual person and capability. In acting thus within a larger whole of rootedness in history and locale, the architect serves in the emergence of architecture the product, of something that is far from a mere thing that sits there mute and alien, but which instead embodies and corresponds to elements of profound significance and impact that reach deep into the very substance of human life and being.

2. The World in the Total Substance of Architecture.

One well-known and widely read author of the history of architecture, Spiro Kostof, defines architecture, in part, as the act, the gift, and the art of making places for some human purpose.[3] But Kostof is quick to point out that architecture is more than just material arranged for some functional purpose, it is also a palpable composition of *images* to be seen by those who use it.[4] We shall turn to this aspect of the necessary *visibility* of the architectural work later; here the matter I wish to dwell on is the way in which architecture is the *defining of places*. For this is the first, and perhaps the most important, of the elements that, of architecture, are more than the human individual, whether user of some building or author of its conception.

Place may in fact be said to define human being. For man is the only animal that looks upon the entire universe from within a perspective and framework that this animal determines for itself. Nietzsche calls man the "not yet determined animal;"[5] but that is not because the human animal never does get determined, but rather because this animal only becomes fully determined, that is, only because actually human, when it places itself within its own schema of determination. To put it another way, man the animal becomes man the human when, in having a culture, it has the framework for its self-location in the universe and for its self-interpretation in terms of that self-assigned place. Here is the whole point of the *myth*, the comprehensive view of how the universe originated and is organized so that *places* are given for all things, especially, and most importantly, for the mythmakers themselves, for man.

The setting of places for the universe, however, does not just occur in *ideas*. It also occurs in the arena of the experientially visible and tangible, that is, in space and among things. Mircea Eliade has spelled out for us one of the main features of the space that figures in mythology, or, rather, the space that mythology figures into our experience of the world, namely, that there have to be founding divisions and founding bases of orientation before there can be places for things and a place for man.

> For religious man, space is not homogeneous; he experiences interruptions, breaks in it; some parts of space are qualitatively different from others. "Draw not nigh hither," says the Lord to Moses; "put off thy shoes from off thy feet, for the place whereon thou standest is holy ground" (Exodus, 3,5). There is, then, a sacred space, and hence a strong, significant space; there are other spaces that are not sacred and so are without structure or consistency, amorphous. . .
>
> It must be said at once that the religious experience of the nonhomogeneity of space is a primordial experience, homologizable to a founding of the world . . . For it is the break effected in space that allows the world to be constituted, because it reveals the fixed point, the central axis for all future orientation.[6]

If, now, we try to imagine space and the world *before* and *empty of* any setting in of points or axes of orientation, a totally neutral and homogenous space, we cannot. To reflect on why we cannot will bring us closer to the sense of architecture as worlds in the making.

We cannot imagine totally neutral space because to imagine space is to imagine *oneself experiencing* space. But think, for a moment, what experiencing space is. Space is not something visible or tangible; space is not a determinate thing nor a definite place. It is the place of things and the place of places, which itself, however, is neither thing nor place. To experience space, furthermore, is to experience it *in orientations*, or rather it is *to experience orientations*; for experience is always perceptually oriented. Perceptual experience is visually open more to the forward than to the sides, though the sides are peripherally included. Visual perception differentiates up and down, one side and the other, near and far. Auditory space completely surrounds rather than being more forward or back, but otherwise it shares the other orientation values. These orientations, now, are more *experiential potentials* for generating places rather than the determinants of actual locations; but the distinction gives us a basis for thinking about the main topic here, namely, *the whole of things*, the *world as a whole*. In order for there to be an experience of the world as a whole, as an unplaced "place" differentiated into primary places, it is not enough for one to possess merely these vectors of experiential orientation, these vectors of possible placing because they are indeterminate, shifting and relative; they change with every move I make. The *world*, however, the place of places, must be firm and absolute, else *places* within it will not hold, and I will be lost and without abode in the boundless universe. There must be some kind of all-encompassing framework, and whatever it is that firmly sets that framework of places will be the founding act for a whole world.

Architecture, now, shares deeply in this phenomenon, if it is the act, the gift, and art of making places for human life and action. Architecture differentiates and places. It differentiates activity and rest within the architectural fabric from what goes on outside it; it differentiates what takes place within it in one part from what takes place in another part; it separates the soaring vault or protective ceiling from the solid floor trod under foot—one could go on and on. But the point is that this differentiation is an act that mimes the founding determination by which the myth sets the articulations for a world. It is no accident that the building often stands as symbol for the universe, and yet the matter doesn't rest with that. The interinvolvement of mythic world-articulation and architectural human space-articulation goes further, as I now wish to try to make clear.

We first of all have to reflect further on what "*world*" means. We tend quite naturally to think of the world in terms of the things *in* the world. We think of all the *things* we see around us, we imagine them all together, in one huge assemblage, and that is what we tend to call world, namely, the assemblage of everything there is, the sum-total of everything in the universe. The world is like a big house, containing every object in the whole expanse of the universe—except that it is a house without any real walls! And then we think of space as the empty distance between the different things in this immense, boundless "house."

The trouble with this picture is that it tries to take the world as itself an object, as the object that contains all other objects. But this is to try to conceive the world in terms that just will not fit. The world is not any kind of thing at all. So we try to conceive the world as the place that contains all other places, in terms of a system of setting one point off from another. The world is not set off against any other "world" because it is supposed to be the "place" that contains *all places.* We can, of course, imagine more than one "world," but then each of them is not "the whole of things;" and another world, a truly all-encompassing one, can be thought of as containing all these mutually different, lesser ones.

So the world cannot really be thought of as either a thing or a place. The only terms we might use will be rather abstract ones, and not really satisfactory; but at least they will avoid the naive categorization just criticized. We could think of the world as the unlimited horizon wherein to set boundaries, as the boundless "Wherein" itself for locating things and setting them off from each other and in relationship to each other. Space, then, would be the fullness of this locating and setting off of things. To experience the world and space, then, is not to experience some enormous thing or container of things, but to experience things themselves precisely *as located*, as set off from each other, as filling out the basic orientations of one's perceptual field.

We do this, however, not in the abstract but in the concrete, we experience the world in terms of *familiar* things in *familiar* places—the world is experienced as the *familiar and known surroundings* of our real life, we experience the world, we would say, as *our* world. But it is just in this way that our naive concept of world will show its limitations; because the world as it is experienceable by us, the world taken in terms of our own familiar surroundings, is only *one possible way* for a differentiation and articulation to be set up whereby a world is woven for the human animal.

We have to try to grasp how it is possible for the world to be *experienced differently* than in the way we are familiar with, not differently in respect to the particular places of our surroundings, but differently in the very way surroundings are conceived and places set, a difference in the very *idea* and *experience* of place. A study of human space and human placement in space from pre-historic times to the present shows some remarkable contrasts.[7] Let me attempt, in a brutally compressed over-simplification, to explain some of this contrast, and to show how that both complicates and illuminates our topic. I shall offer this in the form of seven considerations.

1. The world is the "Wherein" of differentiation and ordering, both with respect to things and events around us as experiencing them, and in respect to the way we humans fit in among the things and events that we experience. In order for this differentiation, ordering, and our consequent fitting in to occur, there have to be certain *governing references* in accord with which it be done. But there is more than one way in which governing references can be formed, indeed, *have been* formed. One of these is for the governing references to be *rhythms, cycles, or paths of movement*. Here differentiation and ordering are in terms of basic phenomena of *process*, so that "place" in this economy is an identification carried along by and *within* processes, *in terms of* and *within* great rhythms, cycles, and movements. Thus the sense of world and the articulation of referential value-determinants in a hunter-gatherer cultural economy of experience is one based primarily on the movement of herds, the recurrent turning of the seasons, and the regular growth of plants in the wild (which is the only kind there are; in fact, they are not even "wild" to the hunter-gatherer, for there is no other kind–the "non-wild," the *domestic*–against which to distinguish the "wild"). Human life and experience follow these rhythms, cycles, and movements like motes of iron caught in the lines of force of some vast fluctuating magnetic field, and their whole value and animation derive from being carried in that dynamic field.

2. With the recourse to the extensive settlement, however, another way of forming governing references emerges, namely, in terms of *fixed centers, pivots, and planes*. Differentiation and ordering are done in a stable frame, which, however, can have two variants. One variant, retaining in strength the sense of process just spoken of, sees this fixed framing reference in terms of *vectors, relationships*, and *forces*. The other diminishes this factor of process, and anchors the frame of reference in terms of *determinate substantial things*. We find places fixed, then, in stable sites, and determined there by a network either of vectors and relationships, or of things and thing-assemblages. Here, too, corresponding to these two ways of

determining places, there is the possibility of two variants in the way humans experience their world; for the world, the "Wherein" of differentiation and ordering by these fixed centers, pivots, and planes, can be itself either *dynamic* or *static*.

3. We now have the specific category for our own sense of world, and the one embodied in our works of building, of determining and creating places. Architecture as we know it has the city as its predominant setting, that is, our architecture at heart has its native place in what we know as the city, a complex living structure based on sedentary activity relating to stable things and differentiated in terms of fixed places. The world for architecture as we experience it is primarily a world of things and thing-assemblages and of the places those things occupy, even though functions are carried out, processes go on, and events happen in those places.

4. The meaning of "world," now, the sense the world comes to have as experienced by humans and imaged in their artifacts, must not be thought to be necessarily and exclusively *like our own*, namely, one in which the idea of the world is taken as the totality of places based on stable things and thing-assemblies. And when we confront the possibility of alternate senses of world, then we are confronted with the fact that the world as we experience it is hardly so entirely *stable* and *fixed* like some great frozen thing or aggregate of things. For to think of it this way is to overlook the whole dimension of *temporality* with which our own world too is infused.

5. Take our own experience of the world, for example. We experience the world primarily as the vast totality of things, thing assemblies, and thing places. The spread of these things and places comprises space, and spatiality thus fills the world. But we also experience this totality of space as existing, and changing, *through time*. The world, then, is not just a spatial "Wherein," it is also a *temporal* "Wherein." And so we think of the spatiality of the world with its three dimensions of verticality, horizontality, and depth as held and modified in the inexorable movement of time. Time is spoken of as a fourth dimension, the dimension of coming-to-be and passing away divided and linked by the poised ever-replaced moment of the now. But if we think of time as a fourth dimension along with, but deeply different from the other three, we must keep in mind that all four are

dimensions of the one great "Wherein" of the world. We only have a complete sense of the world when space is supplemented by time.

However, to take time this way as a fourth dimension after these other three, to take time as processual movement infusing the articulate expanse of space, just underscores how our experience of the world and our sense of space *is not the only possible one*. This whole picture can be radically altered. Instead of conceiving time in terms of a kind of movement or flow from place to place, i.e., in terms of space, one can conceive it *in its own terms*. One can conceive it from the first as a *qualitative transformation* of reality in the very moment of being experienced, as a continuous transformation in three qualitative phases: namely, the yet undetermined possibility of the future, the vibrant actuality of the now, the given and done of the past. Space, then, would be a *fourth dimension of time*, a fourth order of being adding to these three phases of temporal happening. Here space is the radial sphere-like spread of actuality in the moment of qualitative transition from future to past. Space is the swing of differentiation that posits one element external to another in the blazing simultaneity of the now. And, once again, it is this whole temporal-spatial "Wherein" which is the world.

Now, there are serious philosophical reasons— metaphysical reasons, if you wish–for taking temporality as the more fundamental factor by which the structuring of such a thing as the world occurs. There are theoretical reasons for the possibility that the experience of the world by actual humans living it could be one in which process in time rather than fixation in space would be the basic order of reference along which to identify place, and to interpret the phenomena of one's concrete existence. I am suggesting, therefore, that the idea of an understanding of the world which takes time as primary, as a continuous three-phased process with space the fourth dimension that necessarily unfolds out of it, provides a perspective for interpreting the world of some other peoples, and of their artifacts, as a world quite different, even deeply different, from our own.

6. And so the world can be conceived and experienced with space and time in differing primacy and predominance, but in ever conjoined unity. Whichever conception one implicitly has, the *world* remains as the all-encompassing "Wherein" of the ever-renewed cresting of time in the vibrant actuality of space. Our architecture, therefore, and any architecture,

has to be seen as permeated with the particular sense of world that is possessed and lived by its creators as a factor that encompasses and sustains them in their work beyond individual choosing and controlling.

7. Let me offer one final thought in this examination of our sense of world, space, and building, namely, regarding the origins of the sense of world as the totality of thing-places, as the container of all things and thing-places. As we have seen, the origin of this idea may indeed lie in the fact that at some point in their long history humans took to locating themselves, if at first only temporarily, in some kind of stable structure, whether naturally occurring (the cave) or constructed by their own hands (the shelter or the hut). With the hut, a dispositive was at hand to suggest that the nature of their place could begin to be thought of in terms of things or fixed points, rather than in terms of processes or rhythms or cycles. If the hut could become the factor by which stability and order in the surroundings of human living is set, such as happens in the city when the hut is multiplied and magnified, then the world could begin to be thought of as itself a great order of such *things* or fixed points, instead of as the "Wherein" for a network of directions and forces and relationships. So that, in the end, the constructed dwelling, the thing which is a house—especially in the *city*—could begin to set the image for our sense of the world, and space would be experienced in analogy to the interior of the house. The world would be, as it were, imagined as a vast house; and so it is no wonder that in turn the house, and more generally, the building, could acquire the symbolic value of miming the experiential sense of the larger "container" which the world would then be taken to be; and the building would become itself a kind of *small world*, a *microcosm* in which man lived and worked.

Here we have, then, *our* sense of place, the sense of space and world that is operative in the architect, who, as the master in charge of building, receives this conception of the world from the history and culture that is the matrix of his or her intellect and imagination. And this it is that can be felt in the building the architect has created, this is the source of one of the most important efficacies that the architectural work can have, namely, that of being something that humans can experience powerfully as a space and a place, as a little world even, as they move up to and around and inside it.

3. Architecture as the Visible Appearance of an Articulated World.

What we have just been explaining is in fact the meaning of the second half of our title: "World(s) in the Making." For there is a world in the making of the architect's work in more than one sense. First of all, there is an already determined sense of world as a whole already given in the architect's own experience by virtue simply of living somewhere in some culture–say, Philadelphia or New York or Ames, Iowa. The world as the ultimate "Wherein" of all places and spaces has been already been set long before the architect appears on the scene and begins to work. The overall sense of space within which the human places for human purposes can be fixed is already long given for any cultural setting. It is the individual shaping of places for particular purposes that individual humans or groups of them can fix and construct for themselves. Or they hire an architect to do it for them. Yet, as we know, this can be done without the architect–and for longer ages on this earth, *was* surely done without the architect, and done very well.[8] There is a world in this sense already implied, and necessarily implied, in the architect's plan and design, or in the architect's realized building.

But there is a second sense in which one can say that there is a world in the architect's building, the making of which it is precisely the architect's job to achieve. For the building designed by the architect, and eventually built by someone, whether the originating architect or not, is a laying out and an integration of places for human purposes. The architect does not originate the need for the places nor the functions that go on in them; the architect does not create the social and environmental site for the construction he or she is designing. But the architect does determine and articulate the whole of places *within* that site for the fulfillment of pre-existent needs and functions. The architect does create the governing, integrating conception that concretely sets the actual places for humans to do things in the specific ensemble to be constructed–the architect does found a kind of world, namely, the specific individual whole of places distinguished as itself a place within the larger "Wherein" of the whole world. Our culturally particular sense of what "world" means–namely, as we saw, the world taken predominantly as an all-inclusive container of fixed things and set places, or as the sum-total of them–is but one possibility for the way humans might think of the world, or have thought of it. And it is, therefore, also only one of several alternatives that might be embodied in human constructions. But specific alternative as it may be, it is *ours*; and as ours it does possess that ultimate experiential significance that we all feel in the word "world," or, better, that we feel when we look out upon as much of the world as we can see at one time—or . . . when we enter into the ambiance of some particular human construction and begin to sense the order of its places and its functions as a "world" of its own, as a "microcosm."

Thus it is, too, that when we see the vestiges of an ancient experience of place, such as at Carnac or Lascaux, we know there was a whole world there to be lived in, there is a world to be *seen* in these vestiges, in these constructions and compositions. But we don't know quite how to *read*, how to *interpret* the world out of them.

What is it, in other words, about architecture that gives its product, the building, a role beyond simply *being* a "place of places," a "container of places for human functions"? The human animal doesn't simply make things for its use, it also makes things *to be seen and recognized and appreciated* as made for some purpose. The human animal makes things in such a way that their purpose, their use, their significance, *shows*. Human artifacts not only *serve*, they also *display*. And they can be made in such a way that the display of what they are as artifacts becomes itself worthy of notice and appreciation precisely *as* the enhanced, displayed *appearance* of the individual character and function of those artifacts.

Le Corbusier's celebrated definition of architecture, I think, implies exactly this. He says of architecture that it is "the correct and grand knowledge-guided play, in light, of conjoined volumes."[9] This is one of two crucial aspects of architecture for Le Corbusier. On the one hand, the volumes of a building are to be let shine forth in the splendor of their form, so that that form can be seen and understood and appreciated; on the other hand, there is the utilitarian service that the surfaces and substance of those volumes perform in relation to human needs and purposes. For that which is thus to be seen and appreciated in its splendid form is, to begin with, a house, a temple, a factory; and that is essential to its appearance.[10]

The question, therefore, is this: how is there an appearance *of the world*, how is the *world* brought to appear in the structures built? To answer the question, one has to apply the distinction given a few moments ago between two senses of world, the "macroworld," if you wish, and the "microworld." The "macroworld" is the world taken in the ultimate, comprehensive sense as the boundless "Wherein" for spatially and temporally locating and distinguishing things and events, which is not itself a thing or an event nor that to which the categories of spatial and temporal "location" at all apply in any sense. On the other hand, one can think of the world as the "microworld," as the "world" of some particular structured ensemble, whether a house, a garden, a neighborhood, a whole city. Any of these things can be approached as composing a "world" of their own, although, for the sake of brevity, we can talk simply of the *building*, and think about the world of the building, namely, the whole of places that it is the architect's job to order and articulate for the purposes and needs possessed by those who will inhabit and do things within the structure eventually built.

Now, quite simply, I think we have to say that the "macroworld," while in no sense identical with this "microworld" of the building, is reflected in it. That is, the sense of the "macroworld" that we have, namely, as a containing whole articulated spatially and temporally in terms predominantly of things and thing-ensembles, is the *sense* of world that we have in regard to the "microworld" of the building. But it is essential here to understand both the *similarity* and the *difference* between the way we experience the "macroworld" and the way we experience this "microworld" of the building.

Consider first the way we experience the "macroworld." I experience the world in this sense not by actually *confronting* such a thing as "the" world presented before me as an object; for, as we have seen, the "macroworld," the all-encompassing "Wherein" of things and places and events, cannot itself be a thing, place, or event, cannot itself be an *object* for our perception and grasp. Rather, we experience the world in experiencing ourselves in some particular place within the entire ensemble of possible places. In experiencing myself concretely, actually, in some specific place, which means in experiencing myself as placed within an articulated whole of places, I experience the world. In other words, I definitely experience the world but only indirectly, as a kind of horizon, not as something ever able to be *located* in front of me like an object or event.

Now this is one of the ways in which we experience the "microworld" too. That is, by actually being *within* a building, by experiencing myself *in* the structured ensemble of places which is the building, I experience the world of the building *as* its world, *as* the whole of its surrounding, articulated order. As I enter the building, I find myself *placed* within it, either to work, or to carry on some transaction, or to meet others, and so forth. But there is another, very different way in which I can be open to the "microworld" of the building, and that is, precisely *as an object before me*, as a structured ensemble before my eyes and open to appreciation by me in its visible self. I can see the building in its mass and volume, in the disposition of light and shadow, in depth and distance, in height and amplitude, in texture and substance; I can *see* and *contemplate* it as "the correct and grand knowledge-guided play, in light, of conjoined volumes," to follow Le Corbusier's phrasing. I can see it as a whole, and in all of its parts, and in the order of these parts. I can see it in its appropriateness or inappropriateness to the needs and purposes it is to serve, or to its surroundings and its setting (that is, in its relationship to its immediate surroundings in the "macroworld"). I can see it in its *appearance* before me, which is quite different from being aware of its as the ensemble that I experience in being placed within it, in dwelling or in doing things within it. In other words, I can place myself outside it, I can place it before me, in order to be able to *see* it as the *appearing* of a place, or of an ensemble of places, *in concrete, embodied, objective form*. And if, even while *in* it, I study it, instead of simply letting myself *be* in it and letting it be experienced by me as the place in which I am, then in a genuine sense I am outside it while being inside it. To put it yet one more way, the building can *be* a place, or it can be *presented* to me as a place, can be *visible* as a place. It is in this way, therefore, that the building can both *be* a world (a "*micro*world"), and be *visible* as a (micro)world.

With these distinctions, now, we are able to pinpoint something special that the architect does, the special contribution to the substance of the building that makes architecture something more than utilitarian, functional planning. For in addition to conceiving the plan, the articulate functional order of the ensemble of places which is the building, it is architecture to design into the structure the *visibility* of that articulate order, to design in the *manifest appearance* of the world of the building.

The building is always a thing, it must always hold and function for certain purposes and certain needs, whether for simple living, or for buying and selling, or for assembling a group of believers, or for countless other purposes. Someone has to design and build it to serve those purposes, and someone can build it to do that and no more. But one can also design and build it so that it also *appear manifestly* as a place for simple living, or for buying or selling, or for assembling believers. It can be made to present the *manifest appearance* of, say, a structure that assures intimacy and comfort, or a structure that promotes the coming and going of trade, or a structure that draws a group of believers together into unity while ordering them according to their responsibilities. In other words, a built thing is architecture when it is built to *make manifest* any or all of those functions to serve which it is thus built as a thing in the first place. And doing this is what makes architecture *art*, in the sense of *fine* art.

Architecture, then, achieves in a building, beyond its functional service, the manifest appearance of something. As we have just seen, the first of these things brought to manifest appearance is the building's basic functional service itself; but there are a number of other things as well, that often take on a far greater importance for the visible presence that the building as architecture offers. There are, for one thing, certain qualities not so much of the building as a fabric of practical purposes, but rather of the functions that go on in it or of the persons that inhabit or possess it. A building can, for example, display and make manifest the triumphant power or wealth of a ruler or a merchant, or the solemnity and dignity of a civic office. Another equally common instance of something that is often made manifest in a building, beyond its structural function, is some *symbolic meaning*, such as, for example, one can see in the soaring height of a church tower, or the brilliant streaming of light into the vaulted interior of a cathedral nave, or the stepped progression of a ziggurat or a pagoda.[11] One could perhaps cover many such things by saying, in more comprehensive terms, that the building as architecture can make manifest, can make visible a *vision*, a view of things that represents an outlook on the world on the part of the individual or social group whose building it is to be. For example, the vision of things on the part of a ruler that would be made manifest in the palace might be one that sees the realm as subject to the ruler's triumphal power, that symbolizes the place of the king as that of a god holding the orb of the universe in his grasp. Or one could simply see the building as making manifest the vision of the architect, the sense of form that for the creative imagination of that architect best announces the order of places for humans to inhabit, or to work in, or to stand in awe of, depending on what the building is for. One could say that what is made manifest is the architect's vision of the order of places that makes up this particular human microcosm being created; and in turn, that microcosm cannot help but embody and concretely express the architect's understanding of the greater world beyond the building, of the macrocosmic "Wherein" that embraces all things and all places . . . including the master builder him/herself.

So we see that there is more at work in the building than the architect as a human creator, or rather the architect is only an architect when there is more at work in the building than the person of the architect. Architecture has roots in the very possibility of experiencing places, and the very possibility of experiencing places is at the core of human existence; architecture is rooted in having a world and being in a world. Indeed, with "architecture archi-tecturing," there are worlds in the making.

Endnotes

[1]Leone Battista Alberti offers the following designation of the architect, properly speaking: "[B]efore I proceed further, it will not be improper to explain what he is that I allow to be an Architect: For it is not a Carpenter or a Joiner that I thus rank with the greatest Masters in other Sciences; the manual Operator being no more than an Instrument to the Architect. Him I call an Architect, who, by sure and wonderful Art and Method, is able, both with Thought and Invention, to devise, and, with Execution, to compleat all those Works, which, by means of the Movement of great Weights, and the Conjunction and Amassment of Bodies, can with the greatest Beauty, be adapted to the Uses of Mankind." *Ten Books of Architecture*, translated into Italian by Cosimo Bartoli and into English by James Leoni (1755; rpt. London: Alec Tiranti, 1955), p. ix.

[2]Baruch Spinoza, *The Ethics and Selected Letters*, trans. Samuel Shirley, ed. Seymour Feldman, (Indianapolis: Hackett, 1982), p. 52. The text quoted is the major part of the Scholium in Proposition 29 of Part I of the *Ethics.*

[3]Spiro Kostof, *A History of Architecture–Settings and Rituals*, New York: Oxford University Press, 1985. This is not strictly a definition Kostof gives as such, but is gleaned from several passages where variants of the one idea are given, viz., p. 15: "the act of making places;" p. 17: "Architecture, in the end, is nothing more and nothing less than the gift of making places for some human purpose:" p. 21: "the act of making places for ritual action."

[4]Op. cit., p. 19.

[5]Friedrich Nietzsche, *Beyond Good and Evil, Prelude to a Philosophy of the Future*, trans. Walter Kaufmann (New York: Vintage, 1966), p. 74. The term if found in section 62, and in the original German reads: "das noch nicht festgestellte Tier."

[6]Mircea Eliade, *The Sacred and the Profane, The Nature of Religion*,

trans. Willard R. Trask, (New York: Harcourt, Brace & World, 1959), pp. 20-21.

[7]Here minimal indication would come from comparing, say, Knossos with Lascaux and the magdelenian hunting camp, across mediating development in such sites as Catal Huyuk, Stonehenge and Carnac. There is no room to undertake that here even minimally.

[8]Bernard Rudofsky, *Architecture without Architects, A Short Introduction to Non-Pedigreed Architecture*, Garden City, New York: Double Day, 1964.

[9]"L'architecture étant le jeu savant, correct et magnifique des volumes assemblés sous la lumière, l'architecte a pour tâche de faire vivre les surfaces qui enveloppent ces volumes, sans que celles-ci, devenues des parasites, dévorent le volume et l'absorbent à leur profit. . ." Le Corbusier, *Vers une architecture*, (1923: rpt. Paris: editions Vincent, Fréal, et Cie., 1966), p. 25.

[10]"Laisser à un volume la splendeur de sa forme sous la lumière mais, d'autre part, approprier la surface à des besognes souvent utilitaires, c'est s'obliger à trouver dans la division imposée de la surface, les *accusatrices*, les *gratrices* de la forme. Autrement dit, une architecture, c'est une maison, temple ou usine." Ibid.

[11]One study of this kind of manifesting of symbolic meaning is given, for example, by E. Baldwin Smith, *Architectural Symbolism of Imperial Rome and the Middle Ages*, Princeton: Princeton: Princeton University Press, 1956.

ON GOING TO CHURCH AND TECHNOLOGY

Carl Mitcham

Abstract: Utilizing an eclectic, narrative description of the experience of going to church, this paper tries to identify the implicitly non- or anti-technological features of such an experience and to assess their weight. The argument, indirectly made, is that in subtle and important ways going to church offers a counter-experience to the kind with which technology typically presents us, and is thus worthy of some attention by those who seek to determine the limits of modern technology.

1. Introduction

The intention of this paper is to speak, in a moderately phenomenological manner, about the experience of going to church and to consider its relation to technology. It is possible for this to appear as a moderately ironic commentary on a conference that will deprive us of the opportunity of actually going to church, though perhaps few would on other occasions take advantage of such opportunities. Since Socrates, those who practice philosophy have not been known for their piety. Yet my thesis, simply stated, is that going to church is in subtle ways opposed to technology and as such deserving of some philosophical respect.

Lifeworld and Technology eds. Timothy Casey and Lester Embree, Washington, D.C.: The Center for Advanced Research in Phenomenology and University Press of America, 1989.

Let me thus attempt a narrative interpretation of the experience of going to church.

One can, of course, go to more than one different kind of church. To use the word "church" rather than, say, "temple" or "synagogue" or "mosque," also already indicates something about the kind of religious activity being attended to. For present purposes, however, I will prescind from the differences between churches and presume, with the support of only occasional eclectic references, that there are certain similarities between the experiences of going to different churches, and to some extent even to synagogue, mosque, temple, etc. My descriptions will also occasionally heighten or emphasize certain aspects of the experience of going to church by appealing to what might be thought rhetorically exaggerated versions of this activity. Certainly any one experience of going to church will not witness with equal intensity the full range of features attributed to it here. By means of such narrative recollections, however, otherwise overlooked aspects of a once pervasive experience may begin to be more fully acknowledged.

2. Preparation for Church

Going to church begins or is at least foretold in the activities, indeed even in the techniques, of getting ready or preparing to go to church. Getting ready generally entails bathing, putting on clothes, tying shoes, etc. In this it initially appears to be no different than getting ready to go off to work or out to play. Yet one customarily puts on better or at least different clothes than when going to work or to play. Work clothes are either common, more durable, or more functional; play clothes often cover the body less and are either more loose fitting or more brightly colored. The preparatory activities and their materials, not to say their technics, thus already begin to tell us something about what will take place in the activity itself. Going to church is not an everyday activity like going to work (which is often thought to be the proper realm of technology); it is exceptional or special, like play (which, although it has techniques, is seldom spoken of as "technological"). But unlike play, going to church calls for some modesty and decorum and entails less vigorousness; it invites a certain seriousness, not wholly unlike work. Clothing intimates, as it were, a pre-understanding or non-discursive appreciation of the distinctiveness of going to church, one that would have us attend to it both as non-work and non-play.

This pre-understanding can be explored in more detail. Consider, for instance, the case of shoes. I remember that as a child I had two pair of shoes: regular and dress. Regular shoes were for school or for play, dress shoes for special occasions. The shoes I wore to church could actually be either pair, provided they were prepared in the proper way. Sometimes I would wear my regular shoes, but only if they had first been carefully cleaned and polished. When I wore my dress shoes, it was not enough that they were already polished, but they also had to be given an extra dusting off; often I would even wash the soles of my shoes before wearing them to church. After washing them and remembering all the things they might have stepped on, I would be particularly careful, in whatever walking was involved in getting to church, to avoid the cracks in the sidewalk, where squishy things and other contaminations might be hidden, as well as grease spots in the church parking lot. This is not to claim that everyone is so obsessive-compulsive about their Sunday shoes—mine was an obsession-compulsion that had, I might add, a touch of playfulness to it; I took some delight in trying to lure my sister into stepping in just those places I wanted to avoid—but only that my behavior manifested a certain pre-understanding of the activity of going to church, one requiring that the associated technics be in one way or another set apart from their alternative uses.

My mother, whose feet did not out-grow shoes the way mine did, actually had shoes for going to church that at any one time (the boundaries shifted over the course of a year or so) were more or less distinct not only from work and from play shoes but also from other dress-up shoes. Indeed, in some religious traditions it is even considered appropriate to remove one's shoes completely when undertaking certain prayers or entering religious structures. The story of Moses being asked to take off his shoes as he approached the burning bush (Exodus 3:5) has special significance in this regard, and may indicate some mild anti-technical bias.

Even when going to church engages the same clothes as on the occasion of, say, attending a concert, the act of putting on those clothes generally differs in subtle but important respects. As a child, again, I was aware that on Sunday mornings (as opposed to Friday or Saturday evenings) my mother put on just slightly less makeup and paid just slightly less attention or, perhaps more accurately, a different kind of attention—to her hair. She combed her hair with a side-long glance, as it were, not with a full face forward determination to look as pretty as possible. Her lipstick was blotted at least one extra time. Certain earrings were appropriate for a night out that were not at all appropriate for church. There was clearly some kind of self-effacement present even in the act of dressing up for church—a point strongly emphasized by those southern Baptist congregations that ban all make-up at church. Moreover, my recollection is that the Sunday morning toilet involved just a shade less emphasis on the mirror and greater reliance on verbal approval from other members of the family.

The time of day of this dressing, the customary morning–in contrast to the evening dressing for a concert or dinner or date–likewise provides some counterpoint. Human beings do not as readily stand out in the natural morning sun as they do in an artificial light of the late afternoon and night. Dressing up of a morning to leave the home and enter a spacious structure likewise entails cosmetic ministrations differentiated from those undertaken prior to setting off for professional pursuits in daytime offices or evening entertainments in artificially lighted halls. A few years ago one of my daughters bought an electric vanity mirror that actually had different settings on it to reflect just such differential lightings.

But before going to church one not only gets dressed, there is also usually a meal to eat. Here again Sunday breakfast tends to be distinct from weekday or Saturday breakfast. This is most obvious when, as in the pre-Vatican II Catholic tradition, a person refrained from eating breakfast at all until after assisting at Mass. But even in the Protestant tradition of my own childhood, Sunday breakfast had a special character. One striking aspect of Sunday was that it was the only day of the week when the whole family could eat breakfast together. On school and/or work days my mother and father ate breakfast together, while my sister and I were more or less on our own later. On Saturday, when everyone got up at different times, breakfast was even more scattered. But on Sunday morning the family was, as it were, re-united, and often over some mildly special fare: pancakes or grapefruit or biscuits or donuts or something else. Since this meal was eaten at least half-dressed for church, it also had a slightly more formal character than any other possible breakfast. Once again the preparation for church set the stage for church itself.

3. From Home to Church

The process of going from home to the church likewise exhibits its distinctive character. Of course, going to church in most communities fulfills certain social–even political–functions as well as specifically religious ones. And these often come to the fore precisely in the passage from home to church. The exchanges of greetings with neighbors, the informal checking to see who is and who is not going to church, certain concerns about acting in accord with (or making sure one's children act in accord with) various conventions of one's congregation, all serve to point up the social character of going to church. But in the technics of the passage from home to church there are as well some subtle indications of a different sort.

Again, some religions are better at emphasizing this than others. Walking to shrines, even barefoot pilgrimages, are characteristic of many religious traditions, including (in some places and on some occasions) the Catholic. More pointedly, in the Orthodox Jewish tradition *halakhah* does not permit a person to be transported from home to synagogue on the sabbath by means of "work." The riding of horses or bicycles or the driving of cars is eschewed in favor of the more directly natural means of walking. The more low-tech the better, apparently. There are likewise reformed anabaptist communities in which cars and trucks are no longer wholly excluded, as they are for instance among the Old Order Amish, but are on Sunday once again exchanged for the horse and buggy.

In central Texas, on an aunt and uncle's farm where I used to spend lots of time during the summer, trucks were used for going to town on weekdays, the car was special for Sunday. There were exceptions, of course, but generally speaking there was a kind of implicit identification between the car as a special form of transportation and going to church. So much was this the case, at least in my mind, that once when we took the car to town during the week, I asked why we were going to church when it wasn't Sunday. I remember being slightly taken aback to be told that we didn't always have to go to church when we drove the car. And all the way to town and back I tried to figure out why I had asked such a stupid question.

In my own suburban family, with its all-pervasive dependency on the automobile and only the automobile, it was nevertheless customary to clean the car on Saturday afternoon to prepare it for Sunday. And, in my own mind again, this was as much a ritual cleansing as any a simple utilitarian practice. (After all, didn't it always rain right after we washed the car?)

4. The Church Structure and Its Furnishings

Having arrived at the church, before the actual beginning of the church service, one can also notice how the building is different than, say, a dining hall or a library. Whole books have, of course, been written about the symbolism of church architecture: the steeple as pointing toward heaven, the altar in the east so that worshipers face the rising sun, the nave and transept in the form of a cross, stained glass windows filtering and transforming light, etc. But at a more immediate level than such more or less cerebral symbolism is a pre-understanding at work in what might be called the common sense or quotidian experience of church artifice.

Immediately when one walks through the doors of many churches one is confronted by a slightly musty odor, the staleness of a building only occasionally occupied (and even then in a quite distinctive, less than active way), sometimes mixed with a residual odor from incense or burning candles. The use of candles, of course, is itself something of a resistance to modern technology, an opposition to change and an adherence to tradition not at all at odds with the musty smell. (Even those contemporary churches that replace votive candles with little electric lights on timers still use candles on the altar.) The church building is also often more or less cold, prompting one to attention. (In the stillness of the empty church I was as a child sure that there lurked strange, not necessarily evil, creatures. This again prompted a certain level of attention, even when entering for common worship.)

As if picking up on this theme, the furniture of pews, in their distinctiveness from both desks for writing and comfortable chairs for relaxing, indicate in their planned uncomfortableness that going to church will call for wakefulness and listening but, in the absence of any surface for writing, that this listening will have to be appropriated in a more inner manner than at school. Pews, too, are not individual but communal, again setting one up, so to speak, to attend and appropriate in a special way. In the earliest churches there were not even pews, forcing worshipers either to stand together or to kneel. Here again, then, there is a kind of pre-understanding present in the furniture of the church, one that complements and extends the getting dressed and eating of the home-bound preparation and the journey from home to the church.

5. The Worship Service Itself

The worship service itself, the going to church properly so-called, develops and elaborates on this pre-understanding of separateness from everyday technics conjoined with a wakeful attending to something that is to be appropriated in distinctive ways. It begins, typically, with music.

The music in church is not, however, just any music, but music of a special kind. To begin with it can be noted that the most characteristic music of the church is neither performed for passive reception (choir performances) nor as an individual production (solo); it is a music that the worshipers as a group or congregation create. In this communal singing there is both an affirmation and a loss of the ego. As I stand to sing the entrance hymn I affirm myself in the presence of others; I am here. As my voice contributes to the congregational chorus I experience a kind of satisfaction in myself, in my own presence and usefulness to or engagement with this activity. But at the very same moment what any one person hears is not (at least if there is good congregational singing) the sound of his or her own vocalization, but that of the group, even of the hymn itself. I am lost in the congregation, not as an alienated being but as a contributor, one who finds my very self raised, as it were, to a different plane. I do not even listen for myself; I listen to hear the hymn and the congregation of which I am a member singing the song of praise that initiates this rite of worship.

The art of singing does, however, turn one's attention toward oneself in another way. The writer, painter, and sculptor all work, using interior skills no doubt, but applied primarily to materials other than themselves; the singer, by contrast, has only his or her own body and its voice on which to concentrate. More even than dance or sport, the discipline of singing is dependent on no knowledge of the external world and leads to no external or material domination, but is based in self-awareness and self-control directed toward a peculiarly inner faculty and to production of a most ephemeral sort–i.e., of sound. If the weaving and unweaving of baskets was for the monk of the desert a peculiarly spiritual discipline, so is the production and immediate dissipation of the song. More than any other artisan the singer is encouraged to develop a detachment from his or her work. Furthermore, the singing work itself, of breathing in and breathing out, of the vibrating and modulating of vocal cords, is another kind of weaving not unlike that taking place in the reeds of the monk's basket. It develops a sensitivity to the interrelations of tensions and relaxations, the variations in life, insight into which constitutes an important aspect of the spiritual experience. The exercise and experience of singing thus manifests on the level of the human body an apt metaphor for experiences in opposition to those of technological work, with its need to focus on an outer reality and to create material forms that maintain their places in the world even after one takes one's hands away from them. The surplus value of song escapes appropriation by either choir master or chorister and remains, at least for a moment, available to provide sustenance to any and all who would attend to it–including those who actually produce it.

Furthermore, when the opening hymn is over we do not immediately sit back down but remain standing, and the gestalt that has been established is prolonged into speech—the peculiarly appropriate poetic speech of the penitential rite:

> I confess to almighty God,
> and to you, my brothers and sisters,
> that I have sinned through my own fault
> in my thoughts and in my words,
> in what I have done, and in what I have failed to do . . . .

—a verbalization reinforced by striking one's breast three times in conjunction with the articulation of successive layers of self-accusation. Precisely the kind of turn toward the self as singer that has been implicit in the singing of a hymn is now applied to ourselves as a whole. We must attend, not to the world, but to ourselves. We are to blame for what has gone wrong, not only in our words, our songs, but even in our thoughts, and even more profoundly in our acts of commission and omission. Yet even here, the work to be done is not simply ours:

> . . . and I ask blessed Mary, ever Virgin,
> all the angels and saints,
> and you, my brothers and sisters,
> to pray for me to the Lord our God.

> Kyrie eleison.
> Christe eleison.
> Kyrie eleison.

We cannot save ourselves. What was initially a turn from the world to ourselves becomes a return through ourselves to another world. This call upon others and God to bring about the inner work that needs to be done is then followed by the Gloria.

> Glory to God in the highest, and peace to his people on earth . . . .

Together the congregation raises the mind and heart to God, placing themselves in the presence of the transcendent. We listen, but we do not hear ourselves; what we hear is something other and deeper than ourselves, of which we ourselves are a part. Then we are asked to close our eyes and bow our heads, affirming once again with the actions of our body the primacy of the aural and the turn toward the self, as the celebrant recites the opening prayer. When, finally, we are seated again, we are not in the same place we were when we entered the church. We are somewhere else.

Interlude: On Church Music

Music, as has often been noted, is the most emotionally engaging of the arts. Plato and Aristotle both describe music as imitating and evoking the passions (cf. *Republic* II and III [esp. 376e-403c]; *Politics* VIII, 5 [esp. 1339b42 ff.]). Certainly music, more than the plastic arts, can occasion happiness or sadness; much more than prose it can command attention and get "stuck in the mind," as when you find yourself repeating over and over the refrain from some song. We hear a song and it calls forth previously associated feelings. Music therapists use song to get through to autistic children and the mentally handicapped and to calm the severely disturbed. Religion, too, being at least as deeply involved with the affective as with the intellectual life, can be expected to have a natural place for music.

But according to Plotinus and a tradition that runs through St. Augustine and Boethius to St. Thomas Aquinas, music is differentially engaged with the emotions. Music can calm the soul, bind the irrational appetites. Even according to psychologists, music is able to promote joy, sadness, love, longing, and calmness–but not, or at least not nearly as easily, anger, fear, jealousy, or envy. There is a mysterious bond between musical pleasure and the highest forms of happiness such that hearing more than sight proves a better metaphor for the beatific vision. One hears the music of the spheres; angels sing and human beings recite the psalms; human beings cannot see God but they can hear his commands which are sometimes described as like music. The Romantic aesthetics of music from Vico to Schopenhauer and Nietzsche expands on this view.

On a more mundane level, however, one can simply note that aural experience has the ability to penetrate the multiple veils of separation in visual experience (subject from object, surface of object from depth of object, clear and distinct from cloudy and confused, near from far, etc.) and to establish its own reference points and criteria of judgment. Like the divine, it surrounds the listener, is not out or over there. Prose strives through poetry to become music. It may well be that primordial speech was music not prose. Music also takes place in time, as does the spiritual life, for which some spatial journey (the plastic image of temporality) is perhaps the most common metaphor. Developmental psychologists have even confirmed the intuition that being able to hear is more crucial than the faculty of sight for becoming human (through time); a child born blind develops more or less normally, whereas one born deaf experiences manifold developmental abnormalities that are surmounted or by-passed only with great difficulty. The church service begins with music and thus immediately introduces the worshiper into a primordial level of human experience. Indeed, if St. Paul is correct that "faith comes from what is heard" (Romans 10:17), this could not be more appropriate.

But the particular form of the religious commitment to the use of music once again discloses certain anti-technical biases. Music an be either the least or the most technological of the arts–acappella or wholly instrumental and even electronic. Traditionally, especially in relation to religion, it has been the least. Not only is church music typically congregational, but the Christian tradition along with most other religious traditions has maintained some form of resistance to or delimitation of instrumental music. In the early Christian period all musical instruments were excluded from the church–a ban which has remained in force in the Orthodox tradition. During the Middle Ages bells and tambourines were gradually introduced into the Latin West for the embellishment of singing, and then in the late medieval period the pipe organ. Yet periodically there have been movements to return to plainsong or Gregorian chant. A 1958 papal instruction on *Sacred Music and the Sacred Liturgy*, for instance, explicitly excludes "electronic" organs from the church–a position maintained by the Vatican II *Constitution on the Sacred Liturgy*. (Of course, this rule, like many others of the Church, is rather unevenly honored.)

6. The Liturgy of the Word

Within the new world elicited by the opening rite comes the reading of Scripture. The congregation is now seated, receptive. The reading consists of passages from the Old Testament and the New–passages which are often interwoven like the themes of a piece of music–punctuated by a communal recitation from the psalms. In Catholic and Orthodox services the worshipers stand for the reading of the gospel. In some Protestant churches the reading from the Bible is called "the lesson" of the day.

The reading of the Word is followed by its pastoral application in the homily or the sermon. Notes must not be taken, but the worshiper is nevertheless called upon to pay attention, indeed to try to understand and to apply the lesson. It is the pastor's or the preacher's job to assist in this understanding and application which is not to proceed by reliance on the techniques of writing appropriate to and indeed mandatory for all secular schooling today. Beginning with copybook and workbook assignments in the earliest grades and proceeding to individualized note taking at the college level, with written memos and letters and proposals being the coin of communication and application in virtually all businesses and certainly all bureaucracies, writing would seem to be the *sine qua non* of any contemporary understanding and application. Yet not even the most "contemporary" churches suggest that parishioners should take notes–the only exception being radio and television preachers who ask their listeners to take down mailing addresses so they can send in donations! By and large this applies even in Sunday schools. (One possible qualification concerns those Sunday schools that emphasize preparing for "Bible Bowl" competitions and related exercises, although even here it is surely a certain aural memory that is being trained.) Consider, then, some of the differences between trying to understand a lesson using written notes and not using written notes.

One obvious implication of not having to take written notes is that if one is to be tested on the lesson from the Bible it will not be by written examination as in secular schools. One will not be required to repeat facts and figures. The testing on what one knows about the Bible, and indeed the testing on how one leads one's life, is fundamentally different from any examination that relies on the written word. To hear and then to understand and apply is different than to hear and then to write down and learn. For one thing, it is necessarily more selective.

This selectivity troubles us when we are in a literal, literate frame of mind. We want to bring tape recorders to church or back with us to the Sermon on the Mount. Otherwise, how can we know for sure what Jesus really said, what really happened. That an oral tradition carried the teaching of the gospel for at least thirty years after the death of Jesus makes us uneasy. We are aware of our own tendency, over much shorter periods of time, to distort and falsify events, to remember selectively what suits our needs and interests, and to re-read even our memories in terms of changing times and circumstances.

Not to appeal to the less transient character of ancient times, however, we can notice our own experiences in church. In many ways it does not particularly matter what is said, the content of the words. The techniques of preaching indicate as much by shunning those visual aids—the blackboard or overhead projector–required by the classroom. The tone of the words, their flow, has its own message. It may even be that beneath the particular words there is a deeper and more fundamental word, a word one can only learn to hear by becoming somewhat tired and bored by the more superficial hearing to which we are so accustomed. In the most traditional Orthodox churches a standing congregation will quietly mill around, visiting side altars, making devotions, during the course of a long worship service. Some contemporary Catholic parishioners have been known to say the rosary quietly to themselves during a homily. Some hesachysts simply repeat over and over again a special mantra, the Jesus Prayer. Transcendental meditators have even gotten scientists to confirm empirically that the saying of a mantra can have psychological and physiological effects of a markedly calming sort. Although speech is important, particular words may not be; perhaps it is the flow or tone that is important, that can lead to a deeper layer of receptivity. I remember as a child being forced to sit still in church and in that very enforced stillness being enabled to think things I did not think at any other time during the week. Holding my father's hand on my lap, and watching the blue veins on the back of his hand yield to pressure from my finger, I remember being regularly flooded with a sense of mortality. My father was a body, he would die, I would die, everything would die. These were not thoughts that were anywhere near consciousness at any other time during the week. The preacher, at the same time, was speaking about sin, and salvation.

Etymology, which often speaks more deeply than conscious reflection, also may have something to say about the meaning of the time of the listening to the sermon. "To hear" a teaching gives forth etymologically into "to obey" that teaching. "Obey" comes from the Latin *ob-* (to, toward) plus *audire* (to hear). To obey is to hear, to be instructed by the words of another. At its deepest level obedience is not so much forcing oneself to do what another says to do in some kind of external manner as it is truly, deeply to hear the more subtle word coming through the voice of another, being transformed thereby, by the power of the words *qua* words, and then acting differently, accordingly. Just as with the hymn, the real exercise during the sermon is not to listen to the preacher but to the teaching for which he is but an instrument, to let oneself begin to be transformed by that teaching, those words, and then to act accordingly. For this it is enough to hear one thing, to catch only one point, or perhaps only part of a point, and to begin to let it sink in. More than once as a child, and even still, it has been experienced as sufficient to have heard some small truth, to have been in a place where I sat still and only tried to listen, and was able to hear at least one thing.

At the conclusion of the sermon the Liturgy of the Word itself concludes with the congregation standing to recite the creed, the profession of faith. It is a kind of practice in obedience, an effort to get the hang of it. Having heard, we proclaim. That, perhaps, is the fundamental understanding and application for which it is useless to take written notes.

7. The Liturgy of the Eucharist

The entrance rite and the Liturgy of the Word—although not all Christian churches use those terms—are basically similar across the Orthodox-Catholic-Protestant spectrum. The structure of the worship is even quite similar in Jewish services and in some other religious traditions. But following the liturgy of the word, Orthodox and Catholic churches introduce another moment in worship, the celebration of the Lord's Supper. Protestant churches—except for the Anglican and Lutheran traditions—do not make this a necessary part of worship each Sunday. It is the Word and its reception that is crucial. But for the Catholic tradition it is the Eucharist, and the reception of the Word made flesh, that is the center of worship, more important even than the reading and hearing of Scripture.

The Eucharist nevertheless is fundamentally continuous with the Liturgy of the Word. In many ways it repeats all the fundamental movements of worship from the entrance rite to the profession of faith. First there is the procession of bringing the bread and wine to the altar followed by the prayer over the gifts. These actions symbolically remove "what earth has given and human hands have made" from us and thus constitute a sacrifice of ourselves not unlike that called for by the choral singing of the opening hymn and the penitential rite. Then, as if repeating the action of penitential rite and the Gloria, the gifts are blessed and given over to God

> for the praise and glory of his name,
> for our good, and the good of all his church.

This is followed by the eucharistic prayer spoken not by all but by the priest. Although not itself a reading from Scripture, it is saturated in scriptural phrases and allusions, and indeed is intended to recall one peculiarly important event from Scripture, the Lord's Supper, and to apply the command given then by Jesus to continue to perform this act "in memory of me." No disciples took notes, but they remembered. The words, the application of the injunction to perform this action, drone on and on, and no one pays much attention, at least not to the individual words. It is what is happening through or behind the words that is important, not the words themselves. Then, just after the crucial re-enactment of Jesus' own breaking of the bread and drinking of the wine, the congregation comes to life again and recites a kind of encapsulated profession of faith:

Christ has died,
Christ is risen,
Christ will come again.

Following an epilogue to the eucharistic prayer the congregation stands and recites another kind of public profession, the Our Father, a further application of the teaching of Jesus. Then comes the communal partaking of the consecrated bread and wine–and then silence. What more is there to say?

8. Leaving Church

Leaving church is much more ambiguous than the entering of church. The concluding rite is typically quite brief. It has nothing to build up to. The message of the service has been, as it were: "Don't just do something, stand there (and listen)." Now we are supposed to stop standing and go back to doing something. But what? Sometimes the concluding rite tries to build up to something, perhaps a social gathering for coffee and fellowship, or to going out and attacking the evils of the world–but, in reality, it cannot help but be a let down. After all, everyone knows that Jesus himself went from the meal we have just celebrated directly to his own crucifixion. Not only is it difficult to go back into the world, it is not even clear that it is appropriate. The function of going to church is to leave the world. The return to the world is necessary but not in itself desirable. (There is even a passage in the *Republic* about this.) There is a blessing and a call to "go in peace to love and serve the Lord," but it is not clear exactly what this means–especially in a high-technology world. It is not at all obvious how one is to apply a teaching of being still and listening and concentrating on one's self in a world of vision-centered action and re-action and power and domination. Going to church certainly raises a question for that world. One leaves the church with that question more or less ill formed in one's heart.

Coda: Beginnings

The perceptive reader may notice that I have placed more emphasis on the beginning than on the middle or the end. This is not wholly accidental.

Beginnings, in their very beginningness, are often more revealing than either middles or ends. Beginnings are the beginnings of something new or different, so that they stand out in some contrast with what they are new or different from. Once that of which they are the beginning has come into being there is no more contrast. There is only the new or the different which is now no longer new or different.

Sunday itself is, in Christian theology, supposed to be a beginning—of the week, of a new way of being in the world, of a new life. But Sunday has been with us so long now it is seldom any longer anything new at all. In order to re-discover its newness it is perhaps helpful to concentrate very closely on its own beginnings, and to try to hear what is happening in the middle and in the end of going to church through the ear of the beginning. One hearing concerns a certain resistance to or deliminting of technology. In a high-tech world with many difficulties associated precisely with its technologicalness, one may perhaps be even philosophically more predisposed to listen to such suggestions than would otherwise be the case. Such a possibility, at least, has been assumed as present in the exercise of this narrative.

THE RELIGIOUS DIALECTIC AND TECHNOLOGY

Frederick Ferré

Abstract: The phenomenology of religious approaches to technology, as shown by Carl Mitcham's reflections, has two tendencies. One is to reject technology as impious; the other is to celebrate technology and use it to the greater glory of the ultimate. These inner polarities reflect larger moments between faith and reason, prophetic and priestly, ascetic and worldly religion which are ideally not beyond reconciliation. This is illustrated by consideration of the technology of an altar candle.

Carl Mitcham's insightful reflections on the technological dimensions of going to church display again, in a fresh and revealing way, some ancient ambivalences that lie deep within religious consciousness. I believe that what Mitcham has shown us is nothing less than the fundamental dialogue between faith and reason, the sacred and the profane.

I

There is in one phase (or strand or pole) of religious consciousness a decidedly anti-technological thrust. This is well identified by Mitcham as the urge to abandon all artifacts: "Take off your shoes!" The ground you are standing on is holy.

Lifeworld and Technology, eds. Timothy Casey and Lester Embree, Washington, D.C.: The Center for Advanced Research in Phenomenology and University Press of America, 1989.

What does this mean, religiously, to the one who at a pre-reflective level simply finds it unseemly, insulting, wrong, to worship with boots on? I suggest that, first, it symbolizes the religious person's sense of proper nakedness before the ultimate. Standing before infinity we can claim no proper dignity of our own. We are "as dust," next to nothing in our own right. To wear shoes is to claim some stature, to sheath one's extremities in pretensions to importance.

The anti-technological sensibility goes further, second, since shoes symbolize more than dignity: they are also protective devices that shield us, through our own human skills, from the environment. In the presence of the ultimate, such protections are not only futile but impious. Would we dare to design artifacts to defend ourselves from our Maker? Such thoughts for this religious posture indicate wilfulness.

Shoes not only give dignity and protection, they also, third, track the grubby residues of daily affairs into the intensely valued precincts of the sacred. Mud, chewing gum, and even worse can be carried from the world into the sanctuary. Take off your shoes, then! The trash-bearing artifacts of commerce have no place in the holy of holies.

All this comes from a clearly recognizable moment or pulse in religious sensibility. It is the side of religion that stresses the radical break in continuity between the sacred and the profane. It is the thunderous "either/or" of infinite *or* finite, Jerusalem *or* Athens, Christ *or* culture, faith *or* reason. "Choose ye this day whom ye shall serve," echoes in this religious pole. "One cannot serve both God and mammon."

II

Still, in its purity, this first, anti-technological theme in religious consciousness cannot possibly have the only say. There is also an inevitably pro-technological moment in religion. The alternative, an entirely artifact-free expression of spirituality, simply cannot be sustained.

There is the sanctuary itself as artifact. Even the simple, goat-skin protector of the arc in pre-Davidic Hebrew religion was crafted. So also was the arc itself. Building technologies, from ziggurats through Gothic cathedrals to contemporary cantilevered worship centers, have always been employed, stimulated, and stretched by religious yearnings.

Technology, at its root only the implementation of intelligence in artifacts, is simply one of the expressions of human existence wherever and *why*ever humanity is engaged. Crafting a sacrificial chalice is no less appropriate to religious sensibility than is removing shoes. Just as there is no non-technological human life, so there is no non-technological religion.

The second movement or pulse in religious consciousness, then, is the affirmation of intelligence and skill as legitimate servant of the ultimate. Witnessed in many ages and in many ways, this religious pole embraces culture, the arts, science, and the world. Religious architecture, religious scholarship, religious wealth and power should outshine religion's cultured despisers. From wine-making to book-production, technology in the hands of religious persons should be better than the best. In this way will the praises of the first Maker best be given.

III

These are familiar contrasts. Ascetic religion *versus* worldly religion; prophetic religion *versus* priestly religion; outsiders *versus* insiders; faith *versus* reason. What Mitcham's meditation on technology and church-going reminds us most importantly is that beyond all these oppositions there may be a moderate, dialectical synthesis *in concreto*, through the artifacts chosen by pre-reflective faith to embody faithful practice.

The examples run throughout Mitcham's presentation, but for purposes of illustration we may choose to reflect on the altar candle. What does this religiously acceptable artifact symbolize?

I wish to look at it here as the successfully embodied synthesis of faith *and* reason. The candle, first, is not non-technological. It requires intelligent technique for its making. And, like all technologies, candle-making arose in response to human needs. It is not wrong for human beings to desire light in the darkness, any more than it is wrong for humans to desire protection from the elements, the production of food, the means of transportation, the covering of our bodies. It is not wrong, and the candle on the altar shows it, for human beings to use intelligence–whether practical intelligence in the craft traditions or theoretical intelligence in science-led *praxis*–to find means for the satisfaction of those needs.

The altar candle, with its flame, second, can be seen as a simple instrument for sacralizing ancient fire, the most fundamental and most threatening symbol of *Homo faber*, child of Prometheus and recipient of the Titan's dangerous gift. The control of fire by purposeful intelligence is that which most essentially sets apart the human species from all others. On the altar, the candle offers its steady flame in peaceful adoration. Its function is both to give pure light and to let its substance be sacrificed under the willing constraints of worship.

Although undoubtedly technological, third, the altar candle is, as Mitcham observes, not a recent or "high tech" artifact. It symbolizes thereby several important synthesizing values. One such value is the acknowledgement of continuity with the past. Tradition is openly affirmed. The candle is not alien to the fellowship of the saints who have gone before. Another value embraced by the candle is the subordination of the value of efficiency–an electric light bulb would doubtless be more cost-effective per lumens delivered, if that were all that mattered–to other values that matter more. Even the omnipresent electrical grid, great symbol of human prowess, attached as it is to mighty generating plants that are the pride of human technical intelligence, need not always be chosen. The religious sensibility is motivated by other, balancing values.

What I find most important in Mitcham's account, then, is that while there is an anti-technological thrust in religious consciousness on one of its moments, and although, on the other hand, there is an inevitable pro-technological aspect in all human activities, an aspect that is sometimes unreservedly embraced by religious persons, the characteristic stance of the religious practices that he describes points us to fresh appreciation of *modified* technologies that are the fruit of human intelligence in the service of a rich palette of vital concerns. They are the technologies, like "go-to-meeting shoes," that are handled with special care and attention; or like altar candles that symbolize human control of fire within constraints of commitments that transcend economic imperatives.

There are many temptations to "either/or" attitudes in human life, and nowhere are they more evident than in religious sensibility. Since religious values are our most intense ones, the appearance of zealotry should not be surprising. But since religious values are also properly our most comprehensive ones, a religious warning against the limits of zealotry may also be anticipated. Neither Athens alone, nor Jerusalem alone, informed the rich heritage of historic Christianity. Both the Greek *and* the Jewish traditions are woven into that tapestry. Neither faith alone, nor bare reason, is adequate for a religion that would provide fullness and wholeness to life. Both faith *and* reason are required for religious thinking. And, similarly, neither anti-technology nor pro-technology attitudes come from a deep study of religious phenomenology. Instead of the "either/or" of intense but narrow thinkers, we find that the Augustinian "both/and" still guides us well in the technological age.

DESIGNING EXCELLENCE: SOME FUNCTIONAL AND AESTHETIC CONSIDERATIONS

Timothy Casey

Abstract: The crisis in modern architecture regarding form and function is part of a larger cultural divorce between beauty and use, meaning and utility. This divorce, in turn, rests upon the modern technological inversion of the traditional priority of use over production and the user over the producer. Attempts to deal with this crisis have thus focused on a renewed and richer awareness of practical life and its significance for the creation of beautiful architectural forms.

> I think that comfort is a function of whether you think a chair is good-looking or not.
>
> Philip Johnson

Of the myriad topics that fall under the rubric of technology, architecture seems the one field of study most suitable and accessible to philosophical reflection. Indeed, there exists a special affinity between philosophy and architecture. What I have in mind is not the hybrid character of architecture as both an engineering technology and a fine art, but rather a

Lifeworld and Technology, eds. Timothy Casey and Lester Embree, Washington, D.C.: The Center for Advanced Research in Phenomenology and University Press of America, 1989.

more profound trait that includes but goes beyond this aesthetic connection. If we take building to be the "art of making places"[1] that serve as the setting or framework for all human activities, then architects, like philosophers, must always come back to the same fundamental question: namely, how should we live? And, in their own distinctive way, they must also provide some kind of answer to this question, although they often attempt to suppress or ignore it by a variety of strategems, some of which they also have in common with philosophers.

Now the intention of these remarks is not so much to justify a philosophical interest in architecture and its history as it is to address what many consider to be the failure of much (though not all) of what goes under the name of modern architecture. The cold, impersonal, anonymous glass blocks and concrete slabs that litter our cities and increasingly our countryside; the reversal of the socialist ideology of the International Style into the most salient social symbol of the corporate state; the common, almost daily, encounters with so-called "functionalist" buildings that fail to work on even the most practical, mechanical level; the outrageous excuses for dwellings that pass for public housing—all this has of course been discussed ad nauseam. What is not so clear is how we came to this impasse and what can be done about it. Why, we might ask, have architects ceased to ask the ethical and metaphysical questions that were part and parcel of their trade until the nineteenth century? And why, in particular, has contemporary architecture failed in most instances to provide those dramatic backdrops and spaces that situate humans in their world in a meaningful and sometimes even remarkably beautiful way?

Part of the answer, I will argue here, lies in the modern divorce of beauty and function, that is, in that peculiar modern contention that interprets culture as a kind of aesthetic refinement that serves as the icing or cherry on top of the economic-technological cake we are all so busy baking. In architectural terms, this separation and hierarchical division expresses itself in the unresolved dichotomies of architect and engineer, form and function, meaning and use. It is, first of all, important to understand how this divorce came about in architectural theory and practice, and this is what I propose to take up in the first section of my paper. Having set the historical stage, I will then sketch a more intimate connection between architectural beauty and function as a way of articulating architecture's age-old link with philosophy and those concerns that are integral to successful design and building.

1. The Modern Roots of Modernism

The compartmentalization of beauty and function is today a part of our everyday experience. We have for the most part removed the beautiful from that experience and, what is more puzzling, no longer expect to find it there. Aesthetic objects, we now believe, are only to be encountered in such out of the way places as museums, natural parks, and mummified urban settings like Williamsburg, Va. or Chatauqua, N.Y. The everyday world, on the other hand, the world we live in, is a functional world where business is transacted and problems are solved. This world is a relatively new world, for purely functional or utilitarian objects did not exist as a cultural force until the nineteenth century and the rise of the cult of efficiency and productivity. In the architectural sphere, the first inkling of this transformation can be detected in the writings of the seventeenth century French architect, Claude Perrault, who, in attempting to reconstitute architectural theory as an *ars fabricandi*, or technique of a priori mathematical rules and operations, effectively reduced architectural beauty to an abstract formality existing apart from function and lacking in any symbolic or expressive content.[2]

The ensuing philosophical debate *among architects themselves* over the relation between metaphysics and mathematics in architectural theory came to focus on the question of beauty and function. Arguing from a tradition that held architectural theory to be essentially a transcendental justification of architectural practice, Etienne-Louis Boullée concluded that architecture "should be in some way like poems; the images that they offer to our senses should excite us with sentiments analogous to the use to which these buildings are consecrated."[3] As this brief but illustrative passage shows, Boullée continued even into the late eighteenth century to cling to the traditional architectural belief that buildings should embody a metaphoric relation between form and function, expressivity and use, that, in other words, architectural meaning depends on a vision of design and building that encompasses more than technological solutions to narrowly conceived technical problems. In sharp contrast, we have the spectacle of Perrault's most important nineteenth century successor, Jacques-Nicholas-

Louis Durand, rejecting the need for any theoretical justification, metaphysical or otherwise, in favor of a thoroughgoing technique that would simplify the art of building by reducing it to a mere technological process. Significantly, it is to the positivistic triumph of Durand's revolutionary approach that we can trace the unhappy tension between so-called purely functional concerns and aesthetic concerns, with the resultant perception of architectural beauty as merely visual and function as primarily mechanical.

We can thus trace this "aesthetic differentiation," to employ Gadamer's apt phrase regarding the removal of artistic beauty from the lifeworld, to what I would call the modern inversion of the traditional priority of use over production. This inversion is first indicated in architecture by Perrault, inasmuch as his attempt to lay out a rule-based system of proportions was consciously aimed at turning architectural theory into an explicit tool of production, i.e., into a theoretical instrument capable of directing and controlling the various techniques of design and construction. Motivated by a Baconian desire to simplify design and avoid the irregularities and inefficiencies of actual building practices, Perrault sought to make architecture "easier" by submitting its procedures to the exact criteria of mathematical formulas. Both architectural theory and praxis were newly conceived in strictly methodological terms and so were "justified," albeit implicitly and without much fanfare, in terms of their ability to facilitate economic and efficient production. Inevitably, this change in focus, as well as in theoretical and technical criteria, led to the conceptualization of use or function in strictly utilitarian terms and to the reduction of lived, qualitative space to the quantitative space of geometry. This homogenization had the advantage of further simplifying the architect's task by making the problems of function and beauty susceptible to universal, mathematical solutions.

In sum, the subordination of use to production involved two distinct moves: first, the transformation of architectural theory as a mode of metaphysical and existential justification of architectural practice into a mathematical instrument put at the service of the practical how; and second, the concomitant rationalization of this practical how through the theoretical planning and programming of all phases of construction. The result, as became more apparent in the nineteenth century, was an "architecture at a distance," i.e., the removal of the architect from any real involvement in the production process (much to the continuing dismay of contractors, builders, and clients). Its basic presupposition can be described as a faith in the sufficiency of knowing *how* to proceed technically speaking, while leaving the question of the *why* to subjective taste and preference. Even more disturbing is the fact that this theoretical distrust of actual building practices was naturally extended to the everyday use of the buildings, so that for the sake of simplicity (=productivity, efficiency, economy) the lifeworld was pared down to a homogenized Cartesian grid, and all those untidy functions that did not quite fit the pre-conceived, rational plan were simply eliminated from consideration. For all intents and purposes, the architect could—and did—remove himself and his architectural concerns from the messiness of the lifeworld and the construction of actual buildings in that world. And he could do this because architectural theory had become a method, "a formal game of combinations, devoid of transcendental intentions"[4] and oblivious to the activities of daily life as they meaningfully relate to architectural concerns of building and design.

The more contemporary and, it must be said, *celebrated* manifestation of these esoteric historical developments is the artistic haughtiness, if not outright hubris, one observes in the architectural world from Frank Lloyd Wright to Philip Johnson. As Spiro Kostof laments, 20th century architects are renowned for taking a "remote, prescriptive, authoritative approach to the design of our environment."[5] From Wright's immovable furniture to Le Corbusier's machines for living and Johnson's "form for form's sake," there is always that "pretence at omniscience,"[6] that scornful indifference to everyday users and builders that culminates in so many technical, functional, and aesthetic disappointments.

Evidence of this alienation can be found at all levels of the architectural enterprise today. To cite but one example on the technical side, there is a growing anxiety in the American building trade about the loss of control over construction to architects, engineers, speculators, and general contractors. As reported by Tracy Kidder in his most recent techno-drama, "Others organize, design, and finance the work from a distance,"[7] while the skilled craftspeople (e.g., carpenters, plumbers, electricians, etc.) are left "to work out the details," that is, to fit the theory to practice and to face the customer when the fit goes awry. What this arrogance and indifference on the part of architects threaten–aside from the livelihood of the builders themselves–is a loss of true *firmitas* or structural stability, for the craft of actual construction is as resistant to formulaic reduction as are beauty and function. Kidder thus confirms what experience would suggest, that the belittling and degradation of builders and craftspeople can only result in a further severance of the connection between thinking and doing and hence in the inevitable undercutting of technical quality and excellence.

What is most important to grasp in this situation is that this architectural hubris is in large part a symptom of that search for prototypical solutions begun by Perrault in his attempts to formulate immutable mathematical rules and an *ars fabricandi* on a theoretical level. The implicit assumption in the work of Le Corbusier and other adherents of the International Style, namely that human needs and requirements are everywhere the same, was not a new development in the history of architecture. Rather, Le Corbusier and others have finally brought to fruition the standardization and mass production of buildings and houses that modern architecture has in fact aimed at since the 17th century.

Unfortunately, the mass produced house, like any other mass produced product, is a standardized structure to which the dweller must adapt his dwelling habits. Thus, contrary to the best of modernist intentions, standardized function follows standardized form. As one critic mordantly describes the International Style approach, "The job of the designer was to identify the 'correct' solution; once it was found, it was up to people to accustom themselves to it."[8] The achievement of the correct, i.e., universal, solution allowed the architect *to impose from without and from on high* patterns of living and use that bear little or no relation to local customs and needs nor that take into account the natural and artificial qualities of the site. In attempting to make a house as serviceable and efficient "as a typewriter," to use Le Corbusier's own simile, the architect was in effect treating people like typewriters, as if messages and ideas could be typed on them but not by them.

2. Form and Function

Yet, in spite of the curious acceptance of this arrogance and its success in dominating the architectural world, it remains an open question whether the Vitruvian elements of use, beauty, and stability really can be formalized in so rigid a fashion as well as whether the technical and human complexities that combine to determine the creation of beautiful and symbolic architectural forms can be reduced to a single conceptual scheme. It also remains open to us to admit that architectural theory as it has evolved in the last 200 years is sadly incompatible with the broader cultural requirements of architecture and so is unable and perhaps even unwilling to deal with the exigencies of use and taste that have gotten buried in the formalisms of functionalism and an abstract aestheticism. For, from an aesthetic point of view, it is relevant to note that the question of "style" in architecture did not arise until the 19th century, when it became an "aesthetic problem" for the architect, whereas previously architectural forms had always arisen out of the culture as both an expression of that culture and a significant addition to it. Moreover, it is now pretty well agreed that the modernist characterization of architectural beauty as the "promise of function" did justice neither to beauty nor to function, reducing both to the merely mechanical or structural.

Clearly, then, the idea of function and its relation to the beautiful have become obscured, if not totally forgotten. Take, for example, the decorative approach of many so-called post-modernist architects who, by adding a flourish here and a historical quotation there, reduce their art to the prettification of rational structures *post constructione*. This weary compromise between taste and function perpetuates the dualism between engineering and architecture, transforms buildings into monuments, and, what is worse, encourages us to judge our environment merely in a visual way. *Architecture as sculpture*: that is the real upshot of functionalism and the New Spirit and hence explains why we are so disaffected with the built environment. What is rightly called "picture-book architecture" evokes a vestigial sense that architectural forms are meant to be lived in and not just looked at from the outside, that architecture is a functional as well as an aesthetic art, and that, although it is without doubt concerned with the creation of beautiful forms, architecture starts out with the primary aim of solving practical problems. Indeed, what really makes it an art is the way it conceives of those problems and integrates its inspirations toward beauty into its technical solutions, so that to separate beauty and function is to undermine the artistic challenge of incorporating *venustas, commoditas*, and *firmitas* into a coherent vision of human life in the here and now. The disappointment one feels with modernism and so many of its post-varieties is not the result of a rejection of technology and its contributions to modern comfort and livability, but is due rather to a dissatisfaction with buildings and spaces that no longer attempt to speak to or allow for a fuller and more diversified range of human experience.

Architectural beauty is a functional beauty in the same way that music written for specific occasions, e.g., weddings, funerals, marches, etc., is best enjoyed and understood when performed in its functional context. Its generation arises from a vision of the lifeworld even while transforming and giving new meaning to a certain aspect of that world. On this interpretation, form does not *follow* function so much as it grows organically out of it in such a way as to mirror its significance for users and spectators alike. In providing a place for a particular human activity, an architectural form aesthetically expresses the meaning of that activity and hence enables a more active participation in it. By heightening the awareness and vitality of the user through a variety of formal effects (shape, color, texture, acoustics, etc.), the architect enlists her artistic capabilities toward the achievement of a life-enhancing, functional end. The prejudice that this is somehow beneath her status as artist ignores the fact that even symbolic meaning and great beauty in architecture have on occasion been derived from the most mundane and practical of origins.[9]

As Lewis Mumford pointed out almost forty years ago,[10] we need an expanded notion of function, a notion that includes but goes beyond the sheer utility of machines and technical devices to those cultural purposes and contexts that give meaning and proper place to the benefits of mechanization and technical progress. Here the concept of function as it is employed in architecture ultimately refers to those ceremonial, social, political, and religious purposes that generate the specific problems to which the architect seeks specific solutions.

The modern American architect, Louis Kahn, expressed a similar understanding of function by eschewing mere expediency and a facile satisfaction of the client. For it was Kahn who courageously urged his brethern to "go beyond the program and probe the nature of human institutions"[11] and who in practice asked the obvious, but explicitly philosophical, questions: what is a school? what is a laboratory? what is a library? Indeed, Kahn's approach is so unique in the history of 20th century architecture–the exception that proves the rule, so to speak–that it is worth pausing to consider it.

According to Kahn, the architect is part builder, part poet, and part philosopher. He writes:

> Institutions are the houses of inspiration. Schools, libraries, laboratories, gymnasiums. The architect, before accepting the dictates of the requested space, considers the inspiration. He questions himself about its nature, what distinguishes one inspiration from another. When he has understood the difference, then he can be in touch with the corresponding form. The form inspires the project.
>
> A work of art is the creation of a life. The architect chooses and composes to translate the institutions of man into environments and spatial relationships. It is art if it responds to the desire and beauty of the institution.[12]

In order to *build*, the architect must therefore be a *questioner*. The kind of questioning engaged in must be such that it can *inspire* the creation of beautiful spaces and environments, notwithstanding the fact that such questioning can in reality never resolve itself into definitive answers. Clearly Kahn was not afraid to stretch himself (and architecture as a creative act) between the real and ideal in such a way that he literally had to make up rules and procedures as he went along in order to move back and forth between these realms, letting them influence and fertilize one another. The Kahnian hermeneutic circle seems to have worked something like this.

Prior to the actual design of a specific space—and hence to any considerations of the human activities intended for that space–the architect thoughtfully arrives at what Kahn calls a "strong idea." For example, in designing an elementary school she would begin by thinking about the meaning of such an institution in the hope of arriving at a stronger notion of its nature and significance. Such thinking, of course, would not occur in a vacuum but would draw upon what Paolo Portoghesi calls "the interpersonal values of society," as well as upon a deeply personal and informed vision of the world.

Such an idea must be "strong," then, if it is to inspire "the corresponding form," which in turn will inspire the concrete project. Thus, it would seem that for Kahn architectural creation is a kind of platonistic emanation that moves unilinearly from a pre-visioned "form" to its instantiation in the spatio-temporal world. But this is not the Kahnian scenario. For although the initial form results from a combination of philosophical contemplation and architectural memory (of geometry and eternal archetypes), it must nevertheless be tested against the concrete, particular functions (e.g., human activities) it is to house and shape in the real world. The original form, if it is not discarded as hopelessly irrelevant, is re-formed (or "deformed," as Kahn's student Robert Venturi puts it) and hence adapted to whatever "forces" and contingencies the situation exhibits.[13]

In short, Kahn's method is to proceed from the whole (form) to the parts (human activities, site, technical possibilities) and back again to a new, "deformed" form, making architectural design a process of "deformation" which tests the form, for instance, of "school" by means of rigorous empirical comparison with the technical, human and environmental demands of *this* school, the one that is to be built. Portoghesi's comment on this hermeneutic process is most apt:

> There is an oscillation from form to function and from function to form, which gradually tends to either ebb or stabilize at a point of equilibrium, whether static or dynamic. This dialectical vision restores to architecture a breadth of vision which the functionalist statute had denied, a freedom which is born out of intellectual tension and out of discipline, not out of the arbitrariness implicit in the mythology of novelty for its own sake. In this sense Kahn restores architecture, first to itself, and then to its history, rescuing it from the deadly embrace of technology and the risky identification of its figurative repertory with elementary geometry and abstract painting.[14]

Furthermore, Portoghesi hails this methodology as an architectural Copernican Revolution that reverses the modernist formula of "form follows function" taken in its most mechanical sense. Form, in other words, is not built up in analytic fashion out of a Cartesian list of clear and distinct functions (i.e., variables) provided by the disciplines of psychology, sociology, et. al. "In this way one no longer expects the function to generate the form, but rather the form . . . to satisfy the function, and in creative fashion, to contribute something more, a redundancy, a tension."[15]

Nonetheless, Kahn's hermeneutics of form can be read in another way, as placing an expanded notion of function at the very center of both the initial development of form out of a "strong idea" (for what is this idea but an essential insight into a function?) and the deformation of the form through comparison with the anticipated functions of the institution as situated in a specific place and time. In the design of a school, both the knowledge of what teaching and learning *are* and an informed understanding of the ways in which such activities can best occur in, say, the South Bronx *circa* 1989, coalesce into a *functional dimension* that guides the architect in deciding when the design can be judged excellent, that is, as enabling the kind of functions that comprise the functional dimension.

At the most general level, analysis of the concept of function as it pertains to architecture reveals several layers of meaning. These are, in order of importance, convenience, ease of use, and suitability to use. *Convenience*, of course, is simply efficiency: whatever saves or simplifies work or effort. *Ease of use*, on the other hand, goes beyond mere convenience in that it implies that a thing is not only not hard to use but is a positive pleasure. Both of these more or less utilitarian senses of function, however, are themselves taken up into the more pregnant notion of *suitability to use*, which signifies what is fitting in a larger cultural sense, the sense to which Kahn is pointing as "beyond the program."[16]

It is only through the consideration of function in this wider sense of cultural purpose–as opposed to mechanical or structural concerns–that architecture is connected up to the larger context of life in which its artifacts are situated. This connection obliges it to come to terms with an array of natural and cultural forces that in large part determine the forms it seeks to generate and construct.[17] Thus, in a more general sense, the purpose or function of architecture has to do with the relative integration of various and quite disparate forces that, through the power of art, converge to provide the backdrop for the lifeworld. It is this harnessing of forces that a handful of architects now refers to as "place-making." To make places is to carve out a geographical area that, in its unique blend of the natural and artificial, situates human beings with respect to the psychological, historical, natural, social, political, technological, and religious forces endemic to a particular time in history and location in the world. To put the matter succinctly, the function of good architecture is to bring humans together with their world, and it does this by making places, that is, by creating sites in which humans can encounter nature and other humans as they go about their daily work and leisure.

3. Beauty and Use

The idea of architecture I am advancing here clearly places the notion of function at its center and thus finds itself in agreement with Gadamer's contention that "A building is never primarily a work of art"[18] (taking "art" here in its fine arts sense). If a building first and foremost presents itself aesthetically, it draws attention to itself; whereas, as a functional work of art, it should be self-effacing in order to provide a framework, a true backdrop, for human activity. On the other hand, this does not mean that form automatically "follows" function, even in its expanded sense, or that aesthetics should be disregarded in favor of some grim existentialist or socialist realist architectural ideology. For one of the ways architecture sets the stage for the human drama is by setting a mood or tone that stimulates specific feelings, thoughts, and images pertinent to that drama. A building or site not only serves a specific human function, it also comments on that function. That, as we have seen, is part of its function. Even Sigfried Giedion, the official spokesman for the International Style, had to admit that most of us "want the buildings that represent [our] social and community life to give more than functional [=utilitarian] fulfillment."[19] What we want, as Giedion reluctantly implies, are buildings and landscapes that give aesthetic expression to our feelings and ideals, that exhibit some sense of history and our place in it, and that foster pride of place and a sensible attachment to groups and institutions.[20]

One might say that such beauty is the "poetry of function", to echo Boullée's vision of the metaphoric nature of architecture. What inspire and help to generate beautiful architectural forms are those lived meanings that animate a culture, giving it coherence and substance, and the desire on the part of the architect to give these meanings material expression in a way that pleases and delights. What is more, the creation of a functionally "appropriate form" is due not only to a genuine vision of the lifeworld but also to the imaginative capacity to enhance that vision. An appropriate form is a form that in harmonizing with a mode of living also serves to lend a deeper harmony to it. Its appropriateness, in other words, is contingent upon the extent to which it both reflects and *adds to* the meaning and vitality of the everyday life that uses it and surrounds it. Gadamer expresses the harmonization of beauty and use thusly:

> If a building is a work of art, then it is not only the artistic solution of a building problem posed by the contexts of purpose and of life to which it originally belongs, but somehow preserves these, so that it is visibly present even though the present manifestation of the original purposes is strange. Something in it points back to the original . . . . Thus architecture, this most 'statuary' of all art forms, shows how secondary 'aesthetic differentiation' is. A building is never primarily a work of art. Its purpose, through which it belongs in the context of life, cannot be separated from itself without its losing some of its reality.[21]

The fact that profound architectural beauty endures through time is thus explained in terms of the capacity of form to preserve the original functions that were once existentially alive in and through the work but are no longer a part of our world. Such beauty can be said to embody a "world" precisely by virtue of the way it points to those functions that in turn point to their worldly context and horizon.[22]

When such a harmonizing of use and beauty occurs, it is common to speak of such buildings and spaces as having "character," a term that common sensically brings together the functional and aesthetic dimensions we perceive as intertwined and mutually supporting and that also hearkens back to the Greek sense of *hexis* as virtue or functional excellence. It is this quality, I would suggest, that we respond to in good architecture if we are concerned with how we should live and where this can best be accomplished.

Endnotes

[1]Spiro Kostof, *A History or Architecture: Settings and Rituals* (New York: Oxford University Press, 1985), p. 15.

[2]See Alberto Perez-Gomez, *Architecture and the Crisis of Modern Science* (Boston: MIT Press, 1983), pp. 32-45.

[3]Ibid., p. 137.

[4]Ibid., p. 304.

[5]Kostof, p. 743.

[6]Ibid.

[7]Tracy Kidder, *House*, (New York: Avon Books, 1985), p. 115.

[8]Witold Rybczynski, *Home: A Short History of an Idea* (New York: Viking, 1986), p. 191.

[9]See Norman F. Carver, Jr., *Italian Hilltowns* (Kalamazoo, Michigan: Documen Press, 1979), p. 25, where Carver observes that "primitive societies seldom set out to erect purely symbolic forms–the symbolism evolved from the original imperatives of providing secure shelter, whether for themselves or their gods. The thatched roof, for example, a widely used practical element in folk architecture, has always posed the difficult problem of securing and sealing the thatch at the roof ridge. Elaborate solutions to this problem provided the opportunity for highly visible symbolic elements distinctive for each locality. Similarly, church bell towers, which began as a practical means to announce the call to prayer, developed into the highly exaggerated towers that now symbolize 'church.'"

[10]See Lewis Mumford, *Art and Technics* (New York: Columbia University Press, 1952).

[11]Kostof, p. 741.

[12]Quoted in Paolo Portoghesi, *After Modern Architecture* (New York: Rizzoli, 1982), p. 77.

[13]Thus the architect not only determines the nature of an institution by noting its *difference* from other institutions, but also submits this form to the *difference* between it and the historical forces to which it is adapted. This is not to deny that in this adaptation a certain violence or "deformation" is also done to the existing situation. It is to recognize that such violence is quite simply unavoidable and, more importantly, is necessary to avoid a greater violence: the "worst violence," as Derrida puts it, a violence of nonarticulation and silence that refuses to pass through the lesser violence of history.

[14]Portoghesi, p. 79.

[15]Ibid.

[16]See Rybczynski for a more detailed discussion of these elements.

[17]See Carver, p. 26: "By force I mean those natural powers, human needs, and cultural demands that influence the ultimate physical shape of the built environment."

[18]Hans-Georg Gadamer, *Truth and Method* (New York: Crossroad, 1984), p. 139.

[19]Kostof, p. 723.

[20]Ibid.

[21]Gadamer, pp. 138-39.

[22]As, for example, the way the nave of a medieval cathedral pulls one toward the apse, embodying forth the religious procession of a lost religious age.

TRANSLATING TECHNOLOGY: FROM ARTIFACT TO HABITAT*

Gayle L. Ormiston

Abstract: However technology is presented or deliberated, it is presupposed in all forms of questioning or in all forms of fabrication. Yet "technology" is not an example of a transcendental category or a fundamental principle or merely one question or problem among others. Technology is comprehensible only as a relay within a series of relays or assemblage of concepts that always transforms itself, translating over and over again our concern with the conditions under which we would like to live. The issue of technology is the creation and weaving of intermediary concepts and practices.

> . . . *tekhnê,* this is the entire–infinitesimae and decisive–*difference, does not arrive.*
>
> –Jacques Derrida,
> *La Carte Postale*

I

Technology is presupposed in and legitimates all questioning.

To advance such a claim runs the risk of asserting an interesting yet obvious and trivial truth. If one comprehends "technology" in its most common sense, as the application of ideas or routine procedures in a

---

Lifeworld and Technology, eds. Timothy Casey and Lester Embree, Washington, D.C.: The Center for Advanced Research in Phenomenology and University Press of America, 1989.

specific context for the purpose of problem solving, or accomplishing certain tasks at a pre-established level of competence or expertise, then to say "technology is presupposed in and legitimates all questioning" is to say nothing new. If one accepts the view represented in the literature which traces the history of technology, the most notable of which is Singer, Holmyard, and Hall's *A History of Technology,*[1] even though its observation remains seemingly obvious, the assertion itself is a fundamental assumption that informs the perspectives of such texts. Technology underlies all human history; it has guided the development, and has developed in concert with, the epochs and episodes registered in our historical and analytical narratives, and continues to do so. The history of humankind–groups, societies, cultures–is a history inscribed in and through the implementation of artifacts, instruments and machinery of some sort, towards the organized control and manipulation of the environment–"nature."[2]

Etymological and lexical deliberations trace histories as well and contribute, in many cases, to the orientation of discourse and narrative. To be sure, a great deal of what is presumed in the contemporary discourse and analysis of technology studies–broadly construed to include the philosophy of technology, the sociology of technology, the politics of technology, and so on—is conditioned by a certain definition of "technology," what may be called the "standard usage," or an already *fixed* identity of technology. For any analysis or discourse to proceed in such a fashion is to do so on the basis of having accepted *already* a concept of technology that is marked off by certain boundaries. Moreover, not only is it to assume that the boundaries can be circumscribed or known, but to announce a certain regulated concept and use of "technology" is to assume that various "technologies" or "technics" share certain characteristic features. Given this set of conditions, one can ask questions *à propos* "technology," similar to those Wittgenstein poses with regards to "games" and "language": ". . . how is the concept of technology bounded? What counts as (a) technology and what no longer does?"[3]

Sometimes, then, an appeal to the etymological derivations of a term or concept will indicate ways for comprehending subtle differences in use and deployment. But are these differences in use not incorporated already into the lexical narrative that traces the different usages in which a term or concept has been rendered? Can one expect to raise anything more or less from such deliberations? The point of posing these questions is to draw attention to two statements, both taken from *A History of Technology*, regarding what "technology should mean," and to allow these questions to guide the current discussion temporarily. Consider the following announcements:

> Etymologically 'technology' should mean the systematic treatment of any thing or subject. In English it is a modern (seventeenth-century) artificial formation invented to designate systematic discourse about the (useful) arts. Not until the nineteenth-century did the term acquire a scientific content and come ultimately to be regarded as almost synonymous with 'applied science'.[4]

> Technology should mean the study of those activities, directed to the satisfaction of human needs, which produce alterations in the material world. In the present work the meaning of the term is extended to include the results of those activities.[5]

To announce *what* "'technology' should mean," or *that* "'technology' should mean" is to yield or to submit to a specific way of questioning, a special form of posing a question, opening an inquiry. The question which can be answered in this fashion is, of course, "What is technology?" It is not an irrelevant question, but most assuredly one that anticipates not merely a response but a certain *kind* of response–one that fashions in advance the kind of response that will satisfy its demand.

It should be understood that the cursory appeal to the etymology of "technology" that appears in the introductory comments of *A History of Technology* (or here) cannot be taken as a final statement about how technology is viewed throughout the analysis which will be developed in that text. Indeed, it must be understood that the appeal itself is made in the spirit of offering a preliminary response to the question "What is technology?" "It would hardly be useful to consider a term the range of which it is the purpose of the work itself to indicate."[6]

Nevertheless, however preliminary the question, however provisional the response, the invocation of an *etymology*–itself a process of tracing, describing, creating a discourse on "technology"–demonstrates how and that technology is already at work in the determination of its meaning(s). That is to say technology prefigures and circumscribes the question of (its) definition. Furthermore, the appeal to an etymology, again however provisionary it may be, demonstrates how the determination that "technology *should mean*" anything at all is not derived from a historical rendering or tracing of its uses and root terms, which one might expect from such a deliberation. Instead, this determination is derived from an antecedent *trans*lation, a crossing-over of thought and discourse from one universe of discourse to another, from the descriptive to the prescriptive. It is derived from a transference of thought, the placement of "all thinking on the subject" in a certain position, that stages the topical and thematic orientation of a discourse on technology. (After all, all translation imports contraband.)

II

To announce that "technology is presupposed in and legitmates all questioning" is to force a shift in the discourse on technology. It is, in an initial and provisional fashion, an attempt to frame a question. By drawing attention to the ways by which technology is determined, to technology as a question, as a medium—a complex network of interlaced paths—for constructing different strategies and ways for questioning the uses and ends toward which technology is deployed, and the values that underlie that deployment, attention is shifted towards the *ubiquity* and *mediacy* of technology. Technology appears at once as an artifact of human existence *and* as the medium by which the habitat of human existence comes to be what it is. To say that "technology is presupposed in and legitimates all questioning" is a recognition of reciprocity, the reciprocity of practice and product—the structuring processes that mediate the relation between artifice and artifact (tool, instrument, machine, etc.). It is, in other words, an attempt to re-think and to think through the concept of technology, and to comprehend how the concept of technology, whatever its determination, mediates and thus becomes the condition for one's comprehension of self and others, and the habitat in which they dwell. Thus, to announce that "technology is . . ." is to create an-other concept, to render an-other translation, which is, in effect, to inaugurate a continued re-thinking of the concept against a background of undefined and indeterminate specifications.

The statement, "technology is presupposed in and legitimates all questioning," forces a shift in the inquiry, or the questioning if you will. To announce that "technology is presupposed in all questioning" places the concept, and our standard use of the term, at stake: the announcement throws into question the boundaries of "technological discourse"—and necessarily so. For once the boundaries of any discourse have been stabilized, fixed in advance by an antecedent translation of the question or understanding of the field, there ceases to be anything at stake. The risk of questioning, the risk of constructing different strategies for investigation, the risk of creating concepts is removed.

As such, technology introduces a question, the boundaries of which remain open always. It introduces the question of creating (or discovering) concepts, where a concept produces an image or images of itself analogous to those images encountered in the cinema.[7] An image defines its own limits, marks out the boundaries of its field (of objects), and orients the background of comprehension. Technology, thus, introduces itself as a question, designating an open field of discursive possibilities, in the creation of concepts. The anticipation of framing a peculiar kind of response that captures the essence of technology remains unfulfilled. As a question, technology forces an on-going exchange of questions, where responses can be translated into other questions and other ways of questioning.[8]

In a rather cryptic and telegraphic fashion, one might say that technology is its own model: a glimpse of itself given in artifacts, framed and fashioned according to the exigencies of a habitat and the visions of its inhabitants.

III

Perhaps it is at this stage that a more direct return to an etymological and lexical reflection on "technology" is appropriate. Indeed, as already indicated, an appeal to the etymological and lexical derivations of a term or concept indicates different ways for comprehending subtle differences in the use and deployment of a term. Here the subtle differences, the differences initiated through a carefully nuanced articulation or inscription, must incorporate the *artifice of usage* itself. Use creates. It orders, in each instance, a different concentration of significance or value, according to the philosophical, political, psychological, theological, and scientific topography in which the term is deployed. Because use involves repetition or reiteration, there is always the possibility of rupture: the conventions which have established the boundaries of a term or the limits of a concept are "by *essence* violable and precarious."[9] And this possibility of rupture is the possibility of *translation*: the creation of intermediary concepts which can comprehend those which remain familiar and obvious. Translation is displacement, the transfiguration of a habitat, the transference of concepts and practices according to the artifice of *technê-logia*.

According to its rendering in *The Oxford English Dictionary*, "technology" stems from the combination of two Greek terms *technê* and *logos–technê-logia*. *Technê* designates art, craft, skill, a way or manner, a means whereby something is created or contrived, devised by art. *Logos* is rendered as discourse, account, definition, and proportion (as an organizational principle). Combined, *technê-logia* is rendered as a discourse or treatise on an art or arts; the systematic treatment of the arts, sometimes the practical arts collectively. More recent entries cite slight alterations and transformations of the term's usage: technology is identified with "weaving, cutting canoes, making rude weapons, and in some places practicing a rude metallurgy" (1869), and later in the nineteenth-century it is linked to "applied science" (*Mechanical World*, 1882).

But now, what does the etymology reveal? Does it disclose anything more or less than what has been indicated already in the normative definitions announced in *A History of Technology*? Perhaps. Does the more direct appeal to the *word* pursued by an etymological and lexical deliberation confirm the normative determinations? In other words, does a more direct appeal to the *word* affirm or negate the assertion that "technology *should mean*"? I would like to suggest that it does both *and* does neither. The reflection on the word "technology" and the history of its many uses and determinations affirms that history: it is embedded in that history, it arises out of and against that background. An etymology of "technology" reveals a history of questioning concerned with technology as it traces the derivations of its uses. By its very nature, the reference to an etymology of "technology" is an attempt to trace antecedent translations, not in an effort to imitate, to copy, or to reclaim a lost and forgotten meaning. Instead, tracing the antecedent uses and renderings of "technology," as I suppose would be the case with any word, is pursued in order to sketch how translation, the reinscription of the term and concept "technology," is a recognition of the difficulty, if not the impossibility, of defining "technology." Translation affirms antecedent renderings, but in doing so it negates or denies their permanence. Translation is not only displacement, but it entails the transfiguration and transformation of what is.

To be sure, an etymology provides a sketch; it is already a trace and a tracing of antecedent translation. Every retracing of this sketch constitutes a translation, another sketch. And in this way it frames a certain image of technology: one that either confirms an already dominant conception or one that forces a refashioning of an old conception. Herein lies the difficulty of definition and translation. All the predicates, all the defining concepts, all the lexical significations, which seem at once to fall under a particular rubric, are conditioned by and products of technology. Like any other term or concept, the value of "technology" is determined by its use in a network of possible substitutions, in a topography marked by the ambiguity of its constitution, what is all too often, and all too carelessly, called a "context." All too often it is forgotten that this topography is inexorably comprised of ethical, political, artistic, literary, psychological, sociological, and scientific interests.

For the purposes of this preliminary reflection, technology is of philosophical interest because it can be inserted into a chain of other equally over-determined terms and concepts where it replaces those terms and comes to be determined by the other terms in that chain, e.g., art, artifice, discourse, craft, cunning, dissimulation . . . skill . . . . And as the etymology indicates, one might say that by definition this list can never be closed. It marks off the medium or the means for revealing what John Dewey calls a "sequential bond" with a "foreseen" future.[10] In the present or current uses of the substitutions which supply this list of terms, the future is designed, marked out, created. In the current uses of these substitutions, the future remains deferred, the product of technological artifice, the fable of so many discourses.

Within this context, then, technology announces its "usefulness"—certain ways of doing, that certain artifacts and technics can be employed in a specific fashion for definite purposes. But the determination of "usefulness" or "practicality," all of what has come to be called "applied," is conditioned by its being a product of use, a product of the performance of (*and* the discourse on) artifice, *techné-logia*.

IV

Perhaps the announcement that opens this questioning of technology's "significations" should be rewritten. Perhaps the statement should read "the *operations* of technology are presupposed by and legitimate all questioning." Why? What difference does the addition of the term "operations" make in the force of the announcement, in its attempt to force a rethinking of technology, in its attempt to frame a line of questioning? It provides another way for indicating how the operations of technology, i.e., art, artifice, creation, translation, mediacy, framing, and so on, are always already at work in whatever investigations or projects we set out to complete. The operations of technology mediate our practical and intellectual activities, and in their performance our comprehension of technology is mediated–altered, designed, displaced, resituated.

Moreover, to speak of "the operations of technology" is to draw attention once more to the assemblage of terms already incorporated into the discussion that indicate the "operations." Technology is not a "thing," even though there are things that go by the name "technology." But even if it is conceived as a "thing," a tool, technology functions always in the realization of possibilities. The operations of technology are "connective operations."[11] The tool is significant *not* in its immediacy as a tool, but in its *mediacy, as a medium* for knowing the future in the present. If technology marks off a medium for revealing a "sequential bond" with a "foreseen" future, then the artifacts and objects of technology do not designate a place in which the coming-to-be of the future will occur. The significance of technological artifice always lies elsewhere . . . in the connective operations and bonds that constitute our habitat.

What I have tried to demonstrate through a series of all too abrupt and fleeting reflections designed to question, to re-think the concept of technology, is that the history of technology cannot be limited to an analytical or historical narrative that traces the development and use of implements, and the circumstances that lead to their formation. Instead, I have tried to show that there is an active and internal alliance between concepts (here the concept of technology), the operations of artifice which condition the creation of concepts, the sensible artifacts through which we frame the future, and the domains in which these concepts are enacted and engaged.

The concept of technology frames our concerns with modes of habitation. Technology carries us, transports us, transforms and transfigures our dwellings; displaces us, and translates (makes possible) the images of the future which inform the "present." If concepts can be likened to the images encountered in the cinema (as suggested above), then, in their projection, not only do these images define their own limits, mark out the boundaries of its field, and orient the background against which we comprehend the image, they do so because they identify what Gilles Deleuze calls an "out-of-field" [*le hors-champ*].[12] The "out-of-field" is not a negation; it "refers to what is neither seen nor understood, but is nevertheless perfectly present."[13]

All concepts–projected images–frame, and "all framing," as Deleuze notes, "determines an out-of-field."[14] Whatever the concept of technology, it fails to comprehend the entire field of possibilities; it cannot comprehend the topography in which it is enacted or inscribed, even though technology refers to what is both included in its framework and that which is out-of-field. Whatever the concept of technology, it cannot comprehend the "technology," the artifice, the discourse, the *technê-logia* which structures it, makes it possible. Just as our concepts of technology, our uses of the term, frame a parcel of this topography for which borders are established, they are part of that parcel. Concepts localize, isolate possibilities, and thus translate the complex scene of habitation into more condensed field of vision. Our concepts overflow: images at one level of representation always intrude upon and determine the facts and images at another level of representation. Concepts actively and internally constitute an assemblage of possibilities, an assemblage of images that always transforms itself, translating over and over again our concern with the conditions under which we would like to live–our habitat. Does this not indicate the operations of technology? Is this not the *issue* of technology–the creation and weaving of intermediary concepts?

Endnotes

*Portions of this essay appear in "Presentation and Issue: Technology and the Creation of Concepts," in *From Artifact to Habitat: Studies in the Critical Engagement of Technology*, Gayle L. Ormiston, ed. (Bethlehem, PA: Lehigh University Press, 1989).

[1]Charles Singer, E.J. Holmyard, and A.R. Hall, *A History of Technology* (Oxford: The Clarendon Press, 1954-1978), 8 Volumes.

[2]*Ibid.*, see especially V. Gordon Childe, "Early Forms of Society," pp. 38-27, Volume I.

[3]Cf. Ludwig Wittgenstein, *Philosophical Investigations*, translated by G.E.M. Anscombe (New York: Macmillan, 1968), part I, section 68 ff.

[4]*Ibid.*, "Preface," by Singer, Holmyard, and Hall, Volume I, p. vii.

[5]*Ibid.*, Childe, "Early Forms of Society," p. 38.

[6]*Ibid.*, "Preface," pp. vi-vii.

[7]Cf. Gilles Deleuze, *Cinema I: L'Image-Mouvement* (Paris: Les Editions Minuit, 1983), pp. 46-61 and 243-289; see the English translation by Hugh Tomlinson and Barbara Habberjam, *Cinema 1: The Movement Image* (Minneapolis: The University of Minnesota Press, 1986), pp. 29-40 and 178-215.

[8]Cf., Martin Heidegger, "The Question Concerning Technology," in *The Question Concerning Technology and Other Essays*, translated by William Lovitt (New York: Harper & Row, 1977), pp. 3-4.

[9]Jacques Derrida, "LIMITED INC. a b c. . . ," translation by Samuel Weber, *Glyph 2: Johns Hopkins Textual Studies*, (1977): 250.

[10]John Dewey, *Experience and Nature* (New York: Dover Publications,

Inc., 1958), pp. 122-23 and p. 152; cf. Martin Heidegger, "The Origin of the Work of Art," in *Poetry, Language, Thought*, translated by Albert Hofstadter (New York: Harper & Row, 1971), p. 59. Heidegger writes: "*technê* signifies neither craft nor art, and not at all the technical in our present-day sense; it never means a kind of practical performance. [. . .] The word *technê* denotes rather a mode of knowing. To know means to have seen, in the widest sense of seeing, which means to apprehend what is present as such."

[11]John Dewey, *The Quest for Certainty* (New York: G.P. Putnam's Sons, 1929), p. 268.

[12]Deleuze, *Cinema I*, p. 28; English translation, pp. 15-16.

[13]*Ibid.*, p. 28; English translation, p. 16.

[14]*Ibid.*, p. 29; English translation, p. 16.

SARTRE AND TECHNOLOGICAL BEING-IN-THE-WORLD

Thomas R. Flynn

Abstract: Starting with Sartre's "techniques for appropriating the world," I reconstruct and assess his contribution to a philosophy of technology both in Being and Nothingness and especially in the Critique of Dialectical Reason. What emerges is a subtle and suggestive, if inchoate, theory according to which our existential projects are modified by instrumental meaning complexes and are dialectically conditioned by the practico-inert. Technology is found to be "alienating" in a primitive sense but not necessarily oppressive.

In the drawing-rooms of the rich bourgeoisie, machines pay visits on one another and manifest their temporary concord.

Critique of Dialectical Reason

Before Heidegger raised the "Question Concerning Technology," he had already underscored our originally instrumental way of being-in-the-world (the realm of the ready-to-hand, *die Zuhandene*). Human being (*Dasein*) exists in a primarily practical manner; second-order, reflective existence (the domain of the present-at-hand, *die Vorhandene*, the traditionally philosophical world) is a derivative mode. So philosophy, if it is to deal with the most fundamental aspects of our life, *a fortiori* if it is to gain access to the being of beings, must address this prereflective mode–what Husserl

Lifeworld and Technology, eds. Timothy Casey and Lester Embree, Washington, D.C.: The Center for Advanced Research in Phenomenology and University Press of America, 1989.

later termed the "lifeworld" (*die Lebenswelt*).

Sartre takes up this challenge in the existential phenomenology of his masterwork, *Being and Nothingness*, published sixteen years after *Being and Time*. There he develops at least three implications of the Heideggerian thesis of relevance to the question of technology, viz., that human reality (*Dasein*) is individuated by its fundamental *project*, that all knowing is thus primarily *practical* in terms of this project, and that we "appropriate" our world by a set of *techniques* or what Foucault would later call "technologies."[1]

That we are radically "projective" beings is a core assumption of Sartrean anthropology. It implies that we thereby "temporalize" by our surpassing the factical dimension of our situation. It argues likewise that we are never self-identical, that human reality is always "more" than itself (unlike the inert fullness of nonconscious being). This is the ontological source of Sartrean freedom and responsibility. But it calls for the unifying focus in our fundamental "choice" or project. Accordingly, "the *principle* of [existential] psychoanalysis," Sartre informs us, "is that man is a totality and not a collection."[2] In a sequence of brilliant, if sometimes erratic, analyses of literary artists over the next three decades, Sartre attempts to uncover the defining projects of Baudelaire, Genet, Flaubert, Malarmé and himself.

Everything turns on the gathering power of a life project, the way an agent constitutes a circle of meanings/directions (*sens*) in the unity of *a* life. Every aspect of this project is *instrumental* insofar as its meaning (*sens*) hangs on the intention of the agent. The only apparent exception would be the famous "desire to be God," that is, the desire consciously to coincide with oneself, or its "authentic" alternative, the choice to live the creative tension of ongoing nonself-coincidence, the so-called "choice of freedom." But these are not so much "meanings" as the conditions of meanings, not directions but the terminus of any direction. There are no meanings-in-themselves for Sartre; all knowledge, indeed, all consciousness is primarily practical-interested-*engagée*. Thus, one could as well raise the issue of Sartre's relation to philosophical pragmatism as that of Heidegger.[3]

Sartre first broaches the topic of "technology" in a broad sense in *Being and Nothingness* under the rubric of "techniques for appropriating the world." At this point he is already discussing human reality as being-for-other (social being) and is interested in the presence of the *other* that "technique" inevitably implies. Significantly, at this stage "technique" includes language and any form of symbol system. So Sartre's entrance into the discourse of "technology" is through an extremely wide opening, a fact which will save him from the excesses of what has been criticized as the "narrow, engineering usage" of the term.[4] Already, he noted the "objectifying" aspect of techniques of any sort and, to the extent that *Being and Nothingness* conflates objectification and alienation, he finds techniques "alienating" as well. A close examination of these techniques will reveal Sartre's initial, existentialist approach to the question of technology.

It is a well known Sartrean thesis that we are absolutely free, but only *in situation* (BN 509). If we are related to the world primarily in a practical manner, this relationship is mediated by innumerable *instrumental-complexes* that offer a resistance and carry a meaning my free project did not confer on them. This is true of all the cultural objects that populate our world and extends even to language itself. When speaking of my "environment," Sartre underscores the recalcitrance of these complexes, their Bachelardian "coefficient of adversity." Even though my fundamental project actualizes their instrumentality, their "objective" characteristics are not my creation; they belong rather to the "givens" of my situation.

When Sartre's discussion of being-in-situation turns from environment to my "fellowman," the issue shifts from the resistance of instrumental-complexes to the *meanings* they bear. Who says "meaning," for Sartre, says "consciousness." Like Husserl before him, Sartre insists that human reality is meaning-giving (*Sinngebende*). But now it is the *Other* as meaning-giver that concerns him. We find ourselves engaged in an "already meaningful" world that reflects the meanings others have generated. Again, it is cultural objects that come to mind, but this time as directing a (hypothetical) imperative toward us: "Do/use x *if* you want y." Given the practical nature of our situation, all meanings are instrumental in the final analysis for Sartre.[5]

The impersonal, generic nature of these directions both constitute and signify my alienation in the sense of interchangeability. The word in the dictionary is there for the taking like the stop sign or the news broadcast; it is for whoever accepts it. But because I *am* my choices "and the techniques that realize them" (BN 511), the Other enters significantly into the very constitution of my world and my self. In fact, I cannot live in a world inhabited by others without employing techniques. "These techniques are going to determine my belonging to collectivities: to the human race, to the national collectivity, to the professional and to the family group" (BN 512). So if there is an "instrumentalist" dimension to Sartre's approach, the role of the Other assures a "sociological" aspect as well. Already in this "premarxian" work, Sartre acknowledges techniques in our bourgeois manner of decorating our rooms, the French method of skiing, professional jargon and regional dialects. It is via techniques that my situation becomes "mine." But equally, it is precisely techniques that reveal "mine" to have been "theirs." The truly philosophical issue, the ontological question at the base of Sartre's approach to the question of technology and to social theory in general is how "mine/theirs" becomes "ours."

It is characteristic of Sartre's philosophy that such collectivities as nation, profession or class are abstractions the "reality" of which is the datable occurrence of the appropriating act. It is my *act* of surrounding myself with "fashionable" artifacts, of negotiating a turn on the slopes and of choosing a particular word or expression that constitutes my being-in-situation *collectively*. And just as there is no "language that speaks by itself," Sartre insists, so there is no "technique of techniques" that could ultimately obviate the technician (see BN 516-517). This conviction that "we can always make something out of what others have made of us" is the maxim of Sartrean humanism. If the present text stresses the optimistic "we can always . . .," the *Critique* manifests a more acute sensitivity to "what others have made of us."[6]

But such is the ambiguity of "situation" in this earlier work that the line between the given and the taken is blurred. "Freedom," Sartre insists, "implies . . . the existence of an environment to be changed: obstacles to be cleared, tools to be used" (BN 506). And the very act that realizes the means at its disposal equally reveals "the particular potentiality of the instruments" (BN 504). It is this last, of course, which interests us at this stage, for it both underscores the subtle reciprocity between agent and instrument that characterizes all technologies and likewise tempers whatever "angelism" or even Husserlian "idealism" might infect Sartrean existentialism.

What we may combine as instrumental-meaning-complexes, then, affect our projects in a threefold way. First, they limit and direct, resist and enable. Their very mediation qualifies the act and the agent, not just the material artifact. Think of Flaubert's painful search for "*le mot juste*," current controversy over sexist language or (to avoid the impression that this is solely a linguistic matter) the socioeconomic effects on our generation of the development of the microchip. It is examples of the last kind that seem to have raised the question of technology in the first place.

But there is a second way these complexes modify our projects, namely, by thrusting them into the *contingencies* of the material world. My plan to drive to a meeting with a friend may be thwarted by an accident or mechanical failure, a power outage may erase weeks of work at my computer, the cancellation stamp may cover part of my message on a post card. Techniques-instruments confer a vulnerability as well as a power on our undertakings and, since we are our projects, on our very selves. The extent, of course, to which we "appropriate" our world via one technique rather than another may be a function of these likely consequences, but their sheer existence adds an unpredictability to the control we expected through the techniques employed. Consider the frame of mind with which one uses a mechanical typewriter or a word processor during an electric storm.

Not only the contingencies but the *history* of these complexes qualifies our projects. The meanings that surround us are temporalized by the sedimentation of prior interpretative events and practices. A physical object, a cone-shaped hat, for example, may carry quite different meanings in medieval France, in baroque Seville and in the Reconstructionist American South. It is not simply a matter of "projecting" these previous uses onto an indeterminate object as if it were a blank screen. The practice or artifact carries these uses with it, sometimes patently by its style or its name, in a "dead" metaphor, for example, often latently by the uses it has excluded or suggested. Consider "old world" cooking, an English saddle, or the Heimlich manoeuvre.

What is specific to the Sartrean concept of technique is its essential reference to the conduct of the Other. If, *per impossibile*, I lived in a solipsistic world, there would be concrete "doings" but no techniques; "the shoe-repairer who puts a new sole on a shoe does not experience himself as 'in the process of applying a technique,'" Sartre insists, "he simply grasps the situation as demanding this or that action. . . . The for-itself as soon as it assumes a position with respect to the Other, causes techniques to arise in the world as *the conduct of the Other as a transcendence-transcended* [that is, as an objectified indicator of ends]. It is at this moment and at this moment only that there appear in the world–bourgeois and workers, French and Germans, in short men" (BN 521, italics his). It is the "outside" of my means-end pursuit that the Other reveals as techniques: the *how* of my being-in-the-world viewed as Other. What emerges are "objective rules of the conduct-as-object; and these rules become universally valid for all analogous conduct, while the supporters of the conduct or the agent-as-object becomes simply *anybody*" (BN 521).

But true to his existentialist anthropology, Sartre points out that as soon as the For-itself interiorizes a technique "it surpasses the technique toward its own end; it is always beyond the technique which it employs" (BN 523). Not only is the agent the ultimate realization of a technique such that he is responsible for its presence in his world, but that same agent as non-coincidental with his techniques is thus ontologically other than they just as transcendence is other than facticity in any situation. This means, on the one hand, that existential, moral *freedom-responsibility* is preserved in the employment of any technique and, on the other, that technological utopias where non-alienating techniques abound are excluded in principle. There is an insuperable "otherness" to any technique, grounded not only in the duality of consciousness/nonconsciousness that pervades Sartrean thought at this juncture but in the internal negation that characterizes the relations between consciousnesses themselves. The otherness-alienation of technique stems from Sartre's notorious claim that "the essence of the relations between consciousnesses is not the *Mitsein*; it is conflict" (BN 420).

The upshot of this brief summary of Sartre's existentialist theory of technology is that, whether it concern material techniques like machines or "more subtle" techniques like feudalism or capitalism (see BN 522-23), Sartre's subsumption of the ancient sense of *techne* into a philosophy of consciousness reveals the limits and the possibilities of his concept of being-in-situation. The chief limit is the ambiguity that weakens any intentionalist approach to the material world: one senses a reluctance to give matter its due. The most promising possibility is the underlying regard for the ontological and moral primacy of individual consciousness (being-for-itself) in the realm of techniques. If this last is what safeguards Sartrean humanism, it is also what undermines any social ontology that would account for "our world" in any but a psychological sense. What is required is a more adequate social ontology than the looking/looked-at model of Sartre's existentialist classic affords.

With *Search for a Method* and especially the *Critique of Dialectical Reason,*[7] Sartre moves beyond a noetic approach to technology to a dialectically "materialist" one. The limits and possibilities of his existentialist theory are subsumed, not rejected, in the later work.[8] By centering on the Marxian concept of *praxis* rather than on consciousness and by introducing the crucial notion of the *practico-inert* into his discourse, Sartre opens the possibility of a truly social philosophy and finally succeeds in giving matter its due. His treatment of techniques accordingly grows into a more explicit theory of technology.

Individual praxis (roughly, purposive human activity in its material environment) like the consciousness it functionally replaces, is projective and temporalizing. And it is dialectical in a way that existential consciousness was not: it enters into positive relations with other praxes in common projects and into alienating, "serial" relations with other praxes through the mediation of the practico-inert. "Praxis," Sartre insists, "is the real humanity of man" (CDR 156), but the practico-inert is "fundamental sociality" (CDR 318). It is here that the technological finds its dilemma and its ontological home, for the Sartrean problem of technology is the ground problem of every social theory: the basic relation between the individual and the social. That he states it in terms of practico-inert mediation is a function of the peculiar "materialism" to which he now subscribes, a materialism that admits of intentionality and otherness.

The *practico*-inert, as its name implies, is not sheer material passivity; it is the repository of prior praxes, the locus of absorption and deflection of present praxes, the vehicle of counterfinality for future consequences. Though not synonymous, the technological and the practico-inert seem to be extensionally equivalent for Sartre. So whatever Sartre ascribes to the one applies to the other as well. This means that the technological is likewise a "place of violence, obscurity and witchcraft" (CDR 318).[9] The technical will support an intelligibility of its own, for example, but only the analytic reason of cause and effect, of excluded middle and of open and closed switches, not the "dialectical" reason proper to praxis, to life and to history as specifically human. But the technological or "worked matter" (*matière ouvrée*) becomes "Manichaean" and "violent" when qualified as *scarce*. It is scarcity, not matter itself, that gives human history its antagonistic cast, turning praxes against each other and introducing exploitation and oppression into human relations. Accordingly, it is scarcity that makes an otherwise "neutral" technology perniciously alienating.

Four additional concepts must be introduced into the discussion of technology to capture the specificity of Sartre's approach, namely, "process," "exigency," "interest," and "destiny." "Process" denotes the impersonal sequence of events proper to the practico-inert field, a field "full of acts with no author."[10] Such are the steps in colonialist or capitalist development, the "logic" of escalation in the arms race, and the necessities of an ethical system. "In this [practico-inert] field," Sartre explains, "everyone's action disappears, and is replaced by monstrous forces which, in the inertia of the inorganic and of exteriority, retain some power of action and unification combined with a false interiority" (CDR 319). When the kind-hearted employer "has no choice" about paying the lowest wage possible, it is the pinch of the practico-inert process that he feels, his (negative) unity with the competition, the constraints of the market. He is at once the victim and the beneficiary of the process, a typically "serialized" individual.

Practico-inert *exigency* is not limited to socioeconomic processes; there is a specific demand of the material object on the one who works it. In the late eighteenth and nineteenth centuries, Sartre notes, "mines and factories created their capitalists, their technicians, their workers" (CDR 154). "Today," he assures us, "everyone understands, or can understand, how machines, by their structure and functions, determine the nature of their servants as the rigid and imperious future of undetermined individuals and, thereby, *create men*" (CDR 159, italics his). He speaks of "the inert exigency of the object," for example, production schedules, transportation requirements, and the like, which "becomes a mortgage on the future of the new worker, a limitation of his possibilities to which he must submit" (CDR 544-45).[11] Again, in the context of material scarcity, these exigencies are violent: the violence of life in the coal industry, for example, or in colonialist-native relations is a function of the demands of the situation itself, not the immediate product of ill will.

Sartre's concern is to demonstrate how within praxis "there is a dialectical movement and a dialectical relation between action as the negation of matter (in its present organization and on the basis of a future re-organization), and matter as the real, *docile* support of the developing re-organization, as the negation of action." This negation of action shows itself in the form of "objective, negative exigencies" (CDR 159). He cites Lewis Mumford's reference to the steam engine's "requiring" constant care on the part of the stoker and engineer and to steam power's "fostering the tendency" toward large industrial plants as further evidence of the phenomenon of exigency. Of course, the concept of exigency and the feeling we have that makes it plausible are paradoxical: does the machine require the operator or the operator the machine? Sartre, who would have resolved the matter with appeal to psychological *Erlebnis* in his earlier work, now glories in its ambiguity and appeals to dialectical rationality as the proper vehicle for grasping this paradox of technology (see CDR 159-60).

Sartre defines "interest" as "being-outside-oneself-in-a-thing in so far as it conditions praxis as a categorical imperative" (CDR 197). By identifying interest with exigency, he incorporates classical British economics into his dialectical account of practico-inert mediation. Thus private property is the unity of the ownership class, imposing an "external" identity on the in fact atomized individuals whom it maintains in serial impotence. So much in thrall to their interests are the new industrial bourgeoisie, so much the "products of their products," that Sartre can plausibly insist that "machines visit machines" in bourgeois drawing-rooms.

Finally, Sartre employs the dramatic term "destiny" to denote a future inscribed in the practico-inert. He describes it as "an irresistible movement [that] draws or impels the ensemble toward a prefigurative future which realizes itself through it" (CDR 551). Thus the factory worker who procures an abortion because she cannot financially support a child is, in Sartre's words, "carrying out the sentence which has already been passed on her" by her "objective situation" (CDR 235). Since "destiny" is a feature of the collective, the worker's need to abort reflects the exigencies of her paycheck but the destiny of her class. Given the Marxist class analysis that guides Sartre's study of the social, it becomes evident that, for him, interest for the bourgeoisie (the machines imported from England by the nineteenth-century French industrial capitalists, for example) is destiny for the proletariat; the concept of a common interest between socioeconomic classes Sartre holds to be self-contradictory (see CDR 206).

Sartre interrelates and summarizes these terms when he observes that "in so far as praxis is process, goals lose their teleological character.

Without ceasing to be genuine goals, they become destinies" (CDR 663). The necessities of the system, the exigencies of the object embedded in the system, impose a limit to human freedom-praxis that appears to come from the other. But, of course, as *practico*-inert, these necessities are ultimately reflections of human praxes, deviations and counterfinalities, no doubt, but transformations of human action and therefore subject to human negation: "you can always make something out of what others have made of you." What the foregoing analysis brings to our attention is the complexity and the structural necessity that is often involved in the "what has been made of you." A mere noetic conversion, a simple change of heart or project will not suffice to liberate us from "this place of violence, obscurity and witchcraft."

Sartre's reflections on the technological are a gloss on the Marxist ideal of liberating the human from being the product of his product to being his own production. Indeed, large portions of the *Critique* are devoted to analyses of how the product produces its producer: the worker must labor all night to keep the blast furnaces going, the assembly lineman must respect the rhythm of the machine. Even the mill owner is subject to the "exigencies" of his plant and, for that matter, of the competition too, as Sartre's play, *The Condemned of Altona*, makes abundantly clear. But, of course, these are traditional Marxist claims. What Sartre adds to the technological determinism of historical materialism is the grounding of social division and class conflict on material scarcity (a view from the classical economists but given a dialectical twist) and a sober respect for the power of individual praxis to betray the cause of social emancipation. Whatever the prospect for technological advance, socioeconomic liberation is not inevitable.

If, then, we take "technology" in its broad Sartrean sense to denote the entire domain of worked matter, that is, all practico-inert objects, processes and instrumentalities, including the more "subtle techniques" of capitalist or colonialist systems and even the natural languages, we must conclude that technology is Janus-faced for Sartre. It is alienating in the primitive sense of objectifying and entailing a certain otherness wherever it mediates. But this "primitive" alienation (CDR 124) need not be oppressive or exploitative. These features, as we noted, enter history with material scarcity. There are moments within intense group activity, for example, when sworn fidelity, a kind of self-imposed inertia, generates a "free alterity" that enhances and guarantees group praxis as common. At last, "theirs" and "mine" can become "ours."[12]

Not that this primitive alienation is negligible. "It is perfectly possible," Sartre allows, "to conceive of a dialectical *praxis*, or even of labor, without scarcity. . . . In that case, the inversion of the unity of human multiplicities through the counter-finalities of matter would still necessarily subsist. For this unity is linked to labor as to the original dialectic" (CDR 124). Some of the most striking passages in the *Critique* are devoted to analyses of the counter-finalities that undermine agents whose interest lay in the practico-inert. Thus the hoarding of colonial gold in Spain led to the devaluing of her currency through inflation; likewise, the Chinese peasants' project of massive deforestation decreased through flooding the very arable land they employed that measure to obtain. In each case the context is dialectical; the agent by doing what he wants to do puts into motion a series of consequences that counteract his original project. "History, at this level," Sartre warns, "has a fearful and depressing meaning: it appears as though what unites men is an inert demonic negation, taking away their substance, that is to say, their labor, and turning it against all men in the form of *active inertia* and totalization by extermination" (CDR 123). Extended analyses of gold-hoarding and deforestation focus on "the type of action that materiality as such exerts on man and his History in returning a stolen praxis to man in the form of a counter-finality." Sartre separates himself from economic determinists when he adds: "The point must be emphasized: History is more complex than some kinds of simplistic Marxism suppose; man has to struggle not only against nature, and against the social environment which has produced him, and against other men, but also against his own action as it becomes other" (CDR 124).

What would end with the advent of material abundance is not the "witchcraft" of stolen praxis and counter-finality but history as we know it, the conflict that has given our story its Hobbesian hue. For human history has in fact been a tale of oppression and exploitation, of negative reciprocities in a field of material scarcity.

It is *violence*, defined as "interiorized scarcity," that the insufficiency of material goods introduces into the practico-inert field. This phenomenon is evident in such examples as people queuing at a bus stop (more evident still in their failure to queue) or the frenzied motion of the stock exchange. But Sartre wishes to extend this qualification to all aspects of the practico-inert (which we have seen to be coextensive with the technological broadly defined). Thus there is a violence in the language of racism, based on the scarcity of exclusion, in the market economy with its "free" labor contracts and its profit motive, and in the offices and roles that certain socioeconomic systems create such as public censor or colonial tax farmer. Again, Sartre's existentialist commitment to praxis comes to the fore: "the only conceivable violence is that of freedom against freedom through the mediation of inorganic matter" (CDR 736). These forms of violence extend from the mystifications of purveyors of public prejudice to the inertia of the "social contract."

And yet Sartre is as aware as was Marx that technological progress is requisite for the "socialism of abundance" which alone can free us from the ravages of material scarcity (*L'Idiot* 3:189). Although he does not raise the issue explicitly, it seems clear that Sartre cannot be opposed to technology (the practico-inert) as such. Rather, it is the violence, oppression and exploitation that scarcity generates which he condemns. As he points out:

> Goods, products, etc., have a double character in relation to man. On the one hand, they are scarce; on the other hand, this particular product, for example, is a real and present being, which I produce, possess, consume, etc. And no doubt scarcity dwells in it as a negative being through the very precautions I take in using it. But to the extent that I produce it and make use of it I also have relations with it by which I transcend it towards my own ends, by which my praxis is its negation; relations by which the result inscribed in it is, within man's univocal bond of interiority with Nature, a positive acquisition. The creation of a tool or an object of consumption diminishes scarcity—by a negation of the negation. It ought, therefore, as such, to relax the tensions of alterity in the group, especially in so far as individual productive labor is social labor, that is, in so far as, whether individual or collective, it increases the wealth of the community. (CDR 151.)

Sartre subscribes to a technological theory of history to the extent that he admits that the practico-inert is "the passive motor of History" (CDR 122). Not only does it constitute "fundamental sociality," as we have seen, but it carries the inertial as well as the enabling forces of society from one generation to the next, acting as an "inert universal memory." As such, then, the practico-inert bears both the "sentence" and what Albert

Borgmann describes as the "promise" of technology."[13]

The very tools that combat scarcity and increase wealth also contribute to *relative* scarcity, namely, "the impossibility that the group in certain circumstances should grow beyond a certain limit without any change in the mode or relations of production (that is to say, scarcity in the form of the discreet liquidation of non-producers in a given society according to certain rules, and *also* as the selection of undernourished producers" (CDR 153 n.). Sartre has in mind the socioeconomic "necessities" by which a society "selects its dead" (CDR 153). Consider political judgments concerning the allocation of scarce medical resources, policy decisions about research monies or the attitude of the general public toward the migrant and the homeless. To respond that people denied access to the latest medical technology are no worse off than before its appearance overlooks the specific nature of the "interworld" which is *totalized* by the sheer existence of these objects. Sartre wishes to show, among other things, "how alienation becomes the rule of objectification in a historical society to the extent that materiality, as the positive presence of worked matter (of the tool), conditions human relations" (CDR 152).

Finally, one can find the beginnings of a "semiotics of technology" in Sartre's work. It would extend our discussion unduly to pursue the issue at length. One would have to analyze, for example, his discussion of how Leconte de Lisle's wearing a monocle symbolically totalized the "neurotic art" forced upon young poets by the socioeconomic contradictions of the Second Empire. The basis of this semiotics is Sartre's claim that the individual is a "signified-signifier" (*un signifié-significant*).[14] Applied to technology explicitly, this means that "the manufactured object proposes itself to men and imposes itself upon them; it defines them and indicates to them how it is to be used. In order to incorporate this set of indications into a general theory of signs, one would have to say that the tool is a *signifier* and that man *here* is a *signified*." Of course, Sartre can be expected to insist that the tool can have meaning only through human labor (the primacy of praxis that grounds his humanism) and to that extent, the instrument merely reflects its user's knowledge and skill. "But precisely because signification takes on the character of materiality," he argues, "it enters into relation with the entire Universe. This means that infinitely many unforeseeable relations are established, through the mediation of social practice, between the matter which absorbs praxis and other materialized significations" (CDR 161). In other words, the features of contingency and historical sedimentation that characterize the practico-inert in general qualify "materialized significations" as well. So "precisely to the extent that the machine signifies the worker as a practico-inert being, deprived of any particular interest (and of all possibility of having one), it also designates him as a general individual, that is, as a class individual" having a common destiny. "And in this case," Sartre insists, "to produce and to signify are clearly the same thing" (CDR 209).

It is via this semiotics of technology that Sartre's earlier instrumental-meaning-complex is absorbed into his later, dialectical discourse. The signifying/producing dimension of the question of technology is particularly apt for phenomenological consideration. In fact, much of Sartre's so-called "literary sociology" in the *Critique* is really dialectical phenomenology, namely, insightful, imaginative (if sometimes imaginary) descriptions linked to appeals to the conditions of the possibility of social ensembles, especially socioeconomic classes. The fact that praxis is productive of "materialized significations" imbues the practico-inert with a meaning beyond its brute facticity; the fact that such significations can be "forces of inertia," like the Spanish ducats, uniting/dividing serialized individuals from the outside in a common destiny, reminds us again that Sartre's theory of technology is "materialist" in an uncommon way.[15]

In many ways a study of Sartre's approach to the question of technology offers an excellent vehicle for surveying his social thought in general. First, he treats the issue in the broader context of the individual-social relationship that constitutes the core of any social philosophy. Secondly, the very nature of technology as both noetic and physical directs us to the underlying ambiguity in Sartre's philosophy overall, namely, the unresolved duality of *spontaneity* and *inertia*, of the taken and the given in any situation. That he finally opts for a dialectical solution is not surprising; but unless the nature of the dialectic(s) is worked out in detail, this solution will strike many as more verbal than real. For one thing, the relation between dialectical and analytical (in the sense of "technological") reasoning has to be examined more closely. How do analytic concepts fit into dialectical discourse such as that of the *Critique*? Is there a properly "dialectical" technology or must the latter be consigned forever to the anti-dialectic of the practico-inert? And so forth.

Because of the context in which Sartre develops his later thought, viz., the proposed theory of History with a Hegelian "H," his discourse on technology is anything but "analytic" or value-free. Neither does it fit comfortably into the categories by which theories of technology are commonly divided.[16] Sartre's existentialist commitment to the ontological and moral primacy of individual consciousness-praxis saves him from the substantialist view. But he does not subscribe to a value-neutral instrumentalist conception either, especially when you consider his remarks about the "witchcraft" and counter-finality of the practico-inert. The third, so-called "pluralist" view of technology, with its cautious inconclusiveness is scarcely typical of the Sartrean approach. Again, he is painfully aware of the ambiguities of the "situation," as we have seen. But this seldom prevents him from diagnosing our radical difficulties or prescribing the proper socioeconomic remedy. In sum, there is more of an existentialist than a Marxian *hope* that pervades Sartre's theory of technology, an optimism chastened by the insecurities of the practico-inert and the inconstancy of human freedom.

I mentioned in passing the thoroughgoing instrumentalism of Sartrean knowledge, its proximity to a generic pragmatism that is currently returning to vogue. The issue was his lack of any ends that were not likewise means in the Deweyan sense. Yet even as Dewey's enthusiasm for "growth" gave focus and finality to his ethics and social theory (perhaps at the price of internal consistency), so Sartre's commitment to "freedom" in a progressively social sense leaves his "pragmatism" in question. But it also opens a space for developing his dialectical reasoning and the marxist-existential theory of technology that he initiates with its help.

Endnotes

[1]On Michel Foucault's "technology of power," see his *Discipline and Punish* (New York: Pantheon, 1977), pp. 215, 233 and 248. On the "art of existence" (*techne tou biou*) in its different forms, see his *Histoire de la sexualite* 3 vols. (Paris: Gallimard, 1976-84), especially vol. 3, *Le Soucie de soi*, pp. 57 ff. Like Sartre's "techniques" but explicitly without their grounding in consciousness, Foucault's technologies comprise practices, discourse, institutions and the like by which one socially relates to an object and constitutes a subject.

[2]Jean-Paul Sartre, *Being and Nothingness* trans. Hazel E. Barnes (New York: Philosophical Library, 1956), p. 568; hereafter cited as BN.

[3]See Mark Okrent, *Heidegger's Pragmatism* (Ithaca, NY: Cornell University Press, 1988).

[4]Carl Mitcham, "Types of Technology," *Research in Philosophy and Technology* 1 (1978), p. 231. While acknowledging that the term is used analogously, Mitcham stipulates a definition of technology as "the human making and using of material artifacts in all forms and aspects" (p. 232). Sartre would extend the term to include the fashioning of well crafted phrases as well.

[5]This is not to say that Sartre lacks any categorical imperative. One such, "choose freedom," forms the fulcrum of his ethical theory throughout its historical ups and downs. Neither does it mean that he subscribes to an "instrumentalist" theory of meaning such as was favored by pragmatists and certain philosophers of science in the last fifty years. Aside from these cautions, the complex matter of Sartrean meaning cannot be pursued here.

[6]The quotation, which occurs throughout Sartre's postwar writings is usually phrased: "A man can always make something out of what is made of him." *Between Marxism and Existentialism* trans. John Mathews (New

York: William Morrow, 1974), p. 35.

[7]Jean-Paul Sartre, *Search for a Method* trans. Hazel E. Barnes (New York: Random House, 1968) and *Critique of Dialectical Reason* trans. Alan Sheridan-Smith (London: New Left Books, 1976), hereafter cited as CDR.

[8]The details of this subsumption have been discussed at length in my *Sartre and Marxist Existentialism. The Test Case of Collective Responsibility* (Chicago: The University of Chicago Press, 1984).

[9]See Robert D. Cumming's insightful essay, "This Place of Violence, Obscurity and Witchcraft," *Political Theory* 7:2 (May, 1979), pp. 181-200.

[10]Sartre, *Search*, pp. 163-64.

[11]On the "triple objective exigency" of "neurotic art" on the part of French society in the second quarter of the nineteenth century, see *L'Idiot de la famille* 3 vols. (Paris: Gallimard, 1971-72), 3, pp. 662-63.

[12]The ontological mechanics of this transformation is subtle and profound. See my "Mediated Reciprocity and the Genius of the Third," in *The Philosophy of Jean-Paul Sartre*, ed., P.A. Schilpp "The Library of Living Philosophers" vol. XVI (La Salle, Ill.: Open Court, 1981), pp. 345-70.

[13]See his excellent study, *Technology and the Character of Contemporary Life* (Chicago: The University of Chicago Press, 1984), pp. 35-40.

[14]See Sartre, *L'Idiot de la Famille*, 3 vols. (Paris: Gallimard, 1971-72), 3, pp. 432-34.

[15]See Hazel E. Barnes, "Sartre as Materialist," in P.A. Schilpp, ed., *The Philosophy of Jean-Paul Sartre*, pp. 661-84.

[16]See Borgmann, *Technology*, p. 9 ff.

THE HIDDEN LIFE OF TECHNOLOGICAL ARTIFACTS

John J. McDermott

Abstract: A denial of the rigid distinction between the organic and inorganic, inert—febrile; the relation between the literal and the symbolical; the dramaturgy of environment as bland, aesthetic, hostile; artificial representation of artifacts; the threat of obsolescence and the consequent antiquing of the recent present; and clinging, nostalgia and sadness as constitutive of the human plight.

The press of my foot to the earth springs a hundred affections. They scorn the best I can do to relate them.

Walt Whitman

Some among us reach for the stars. Still others pine for a revelation, secrets yielded, a private wisdom. For me, I scour obviousness, collecting flecks from the ordinary. In a contentious paraphrase of St. Paul, I offer that it is the world writ small–as in persons, events and things–that is the matrix in and through which we live, move and have our becoming, on the way, undoubtedly, to our personal oblivion.

Of these three moorings in our journey, our trekking to nowhere, *philosophically, things* are the least appreciated. They seem to be more the stuff of art, fiction and technology, as in necessary artifact. Yet, I hold that

---

Lifeworld and Technology, eds. Timothy Casey and Lester Embree, Washington, D.C.: The Center for Advanced Research in Phenomenology and University Press of America, 1989.

it is precisely our things which provide us with our most distinctive, albeit failing grasp on our distinctive human project, to survive 'being-in-the-world.'

1. Things

Things, in the received wisdom, are guilty of two major offenses: one, they are not natural but rather artifactual. (Called by some, incorrectly and deeply so, artificial.) This leads to the alleged claim that things are inert, lifeless and impersonal; two, things are regarded as transient, subject to obsolescence in contrast to the self-deceptive assumption that the person who has/uses things has some permanency.

To the contrary, we hold that things are potentially as febrile as live creatures, even if they are not formal, sanctioned objects of art. Further, it is our belief that things are symbolically pregnant and in fact, following Emerson, often "lose their adhesive, inert form and spring into the empyrean."

2. Natural/Artificial/Artifactual

The pejorative distinctive between the natural and the technological as one between the appropriately human, the worthy and the sensuous as over against the functional, the plodding, and the lifeless is a source of deep distress in our attempt to build a human world. This baleful denial traceable to Plato and to one-half of the dual legacy of Christianity—that things made by humans are genuinely human things, casts us into an unreal relationship with our environment and is more characterized by alienation and amputation than by intimacy and sharing. This situation is more a matter of the absence of sanctioned articulation as to the human continuity of things than it is to an abiding hostility. If we were to approach this bifurcation between our things and ourselves from the back side, that is, from our systemic loneliness, we would find how deeply we need our things.

To *miss* is to be lonely, to be alone, even in a crowd, to be without, to be 'adrift' while yet stationary. It is for us, as in I can't get started, I can't get it together, I am out of place, I am no place, I am nowhere, I am non-body, no-one. As in "something is missing." To miss is to be in a state of hunger, not so much we do not have personal food but rather because the food, for us, is not nutritional, or even it is potentially nutritious yet we do not have the capacity to make the conversion. We eat the world and are not sated. We become more of a viaduct than an organism. That is why we seek *company* etymologically, to be with bread, the stuff, the staff of life, of nutrition. I refer here to flora and fauna, the bugs, the birds, the animals, the ferns, the flowers, the weeds, the water, the water–always the water, on which we frolic, into which we dive, for fun, for mastery, for suicide. The water which feeds us, calms us, frightens us, drowns us–the water as deep, as grave, as home for creatures, strange, mysterious and in temporary recession, but someday to wash over us, to wash us away, to take us back to our first home.

Still, although we do not seem to know it in the same way, systemic loneliness is due also to separation from our things. This separation is *obviously* severe when it is spatial, as in our things are somewhere else, or tragically, when we are ripped from them as we are forced to flee. Yet, more subtly, it is also severe when we are still in the presence of our things, though we do not let them speak to us or, if they speak, we refuse to hear them. I do not underestimate the extraordinary symbolic significance of the social fact that "bag ladies," forerunners of the now pandemic urban homeless, carried their things and had, as a mooring, no place, but only their things.

Indeed, our hearts are restless–yet, for many of us the 'thee' in which we would find rest is elusive. A morphology of suicide is revealing. Some choose violence as being smashed to the concrete in an effort to crush our own spirit, the source of so much pain; or blowing ourselves away with a double-barrelled shotgun (Hemingway) or a magnum 357; or hanging ourselves as an attempt to dangle for all to see and a demand that others, the source of our pain, "cut us down;" or suicide by drowning, a disappearing into the deepest of earth's secrets, only to float up as either flotsam or jetsam, food for the sharks or the piranha reminiscent of the life lived which caused the suicide in the first place.

As with all human experiences, loneliness finds itself on a time-space matrix. Is the *space* I am 'in,' my space as in being someplace, my place or is the space alien space? The same space can be comforting and alien dependent on its peopling (as in a hotel after a convention–the loss of faces, community, common endeavor, bonds assumed without having to be publicly articulated); or when the space is transformed by light or lack thereof, sound, smell; when the space is mobile as in an automobile or a train, constantly visited by shifts, some dramatic, some prosaic, all telling such that my experience of my present space is a flighting, a process, forcing me to claim and reclaim my place in this ever-changing space, despite its apparent stasis. I would extrapolate even further and claim that to be simply in and of the world is to be able to participate in a panoply of shifting spatial contours, facades, depths, alleys, heights, and reaches. Loneliness appears when this panoply fades into a sameness resulting in the irony that the absence of difference structures, not recognition, not habituation in any positive sense, but rather an aloneness, a dried-up desiccated, second-hand person, alone with himself or herself.

The second matrix in and through which we live, move and have our being is *time*, the inevitable source of our erosion and the harbinger of our death, our dismissal from the journey. To escape from this insistent, persistent presence which appears in micro-seconds, we adopt strategies, all of which cut us off from nutrition and, contrary to the salvation we see, we find ourselves in one bunker or another, once again, despite our best effort, lonely. For example, some of us retreat to a vicious nostalgia, whereby the past takes on a glow that it never had as present and seduces us into living our lives backwards. Increasingly the present loses its reality and we become tied to a chimera with our experiences being lived both vicariously and by hindsight. When we are reminded of the unreality of our nostalgia, we become defensive and further enlarge our dependence on a past which never happened and which prevents us experiencing our own present. In effect, we have dropped out and back, isolating us from continuous nutrition and trapping us in a world that cannot be understood or shared by anyone else. We become our own glass menagerie.

Some others of us deny the call of the present by entering into the seduction of a projected future, riven with the deep although most often required nostrum that things will get better. In the meantime, we either bunker down waiting for forces other than our own to propel us or even to rescue us, or we tint every experience with its alleged future-time prospect, in the hope most often vain, that things will (must?) get better. In both strategies, the present has no feet, no past, for that disappears into a redeeming, albeit vague, future. And, further, the power of the present, the existential moment has no bite, for we are always waiting for something else, something different, something better; a wait which is futile, for only the experience of the present can lead to a viable future.

3. Nostalgia and Obsolescence

It comes as no surprise that Western culture has its own version of ying and yang. On the one hand, we cling to the past and retrospectively endow it with a majesty limited only by our hyperbolic abilities. On the other hand, we trash the past, rendering it obsolete in a way that cuts the ground, as source of sustenance and affection, out from under us. The first attitude, known by the rubric of vicious nostalgia, is occasionally troublesome as we find in the person of Willy Loman, the protagonist of *Death of a Salesman*. The second attitude is characterized by the severing of function from memory. When things no longer are 'useful,' we obliterate them, destroy them, abandon them. In so doing, we certainly endanger and perhaps even obviate the memory trigger. The result is a present-mindedness, causal of a banal mind, the absence of oracular examples and very spiritual top soil.

4. A Rejunvenated, Modern Stoicism and Nature as Things, Things as Nature

Now what does this come to for our being here, together under the rubric of still another powerful philosophical method, the phenomenological? Surely, if phenomenology has a contribution, it is its alleged focus on the palpable. The *mal genie* of Descartes does not intrigue me. I cannot, for sure, deny the possibility of a Platonic secret, nor the hints of the too-often arrogant proponents of New Wave esoteric religion. Yet, to put it forthrightly, I am not interested. I embrace the remonstrance of Camus, that I want to live with what I know, and with that alone.

I long ago resonated to the response of James to his own plaintive cry as to how do two minds know one thing? He responds with alacrity and epistemological boldness, they do not!

Surely, it is clear that our epistemological history has its roots in the non-negotiable conflict between Plato and Aristotle. The first great conflict as to the upshot of what it means to be, to know, and to be in the world, is that between Plato, on the one hand, and Aristotle and the Stoics, on the other hand.

For Plato, to be is to be in a trap, a cave, depressingly, in a body. The body as cave. Ascent from the sensual is the way to salvation. To trust the world as sensed, seen, heard, tasted, felt, is to be self-deceived. There is a secret given only to the few. The rest of us are left behind, groveling in an experience which allegedly is a lie. (As if experiences ever lied. Only the conceptual articulation can lie.)

The "way," as in 'I am the way,' that one escapes the cave is appropriately murky. Some say that a voice whispers the secret (Augustine). Others say that the chains of the prisoners are magical and self-will can drop them from us (Norman O. Brown). For still others, the prisoners in the cave, that is, we, you and I, plot an escape. In effect, there is a fight in the cave, a political fight (Marcuse). Yet, as we know from Plato's *Allegory*, those who remain behind refuse to accept the wisdom of the escapee. We are not only *in* a trap, we trap ourselves. The veil of ignorance shines upon us as security, for we do not choose to be, only to accept that we be. We are the deaf in the aporistic sayings of Heraclitus, the Logos speaks but no one hears. The Gospel tells us that we have ears but do not hear, eyes, but do not see.

For Aristotle, there is no secret. The world speaks (in Greek only) and we hear. More, when *we* speak, it is to say how it is with the world. Everything fits and everything, especially the human organism has a natural place. The epistemology of ultimate intelligibility and the metaphysics of coziness, phylic and categorical, assures us of a one-to-one correspondence between reality and person. The sustenance for this approach is not to be understated, namely, common sense or what the analysts derisively call folk psychology.

Phusis as speaking, the Logos, the chattering, singing, murmuring of the eco-system. For Aristotle, we are not alone, we are not bereft. The fish, the rocks, the estuary all talk to us. What do they say? Hey, here it is, how goes it with us? How does it go with you? Similarly, To what end? The telos is cyclic, unredeemed, but not banal. Let us take a moral holiday here, using the phrase of William James and the wisdom of John Dewey.

The human organism, strikingly akin to other advanced organisms, transacts with the affairs of nature. The human organism, however, is paradoxical in that it knows whereof, whereas and whereby this transaction takes place, in short, the human organism is aggressively self-conscious. This transaction between the human organism and nature is experience.

The transaction is ever striated with not only problems, knots, but with the problematic as an ontological condition of being in and of and about the world. To be human is to be constantly, ineluctably, irreducibly faced with the problematic.

For Dewey, one can make one's way through this network of difficulties, although attempted resolutions inevitably generate new difficulties. Life has *no final perch*. Many years ago Dewey described this viewpoint to a lay audience. An elderly woman approached Dewey after the lecture and said to him as follows: "Mr. Dewey, you describe life as though one climbed a mountain to the top and then descended, only to climb another mountain to the top. Mr. Dewey, what happens when there are no more mountains to climb?" Dewey answered in properly laconic Yankee form and true to his metaphysics, "You die, Madam!"

Make no mistake, we have difficulty spelled out in Dewey's version of the human quest. Yet, equivalently, make no mistake, we do not have nihilism. Dewey accepts the claim of James that experiences are indeed cognitive of one another, so that the transaction is not without a guide, a source, a leaning, a thing, a hunch, an Indian head watch. Keep your eye on the ball, it speaks. Listen to the wind, it speaks. Hear the murmurs, read the gestures, mark the gait, catch the tone, eye the color, reach and keep one foot on the ground.

Formal education teaches us that experience can be denotated. The brilliant phyla of Aristotle is a masterpiece of local organization. Modern molecular science has shown that the makeup of nature can be put differently, although still in a conceptual structure. It is Dewey's wisdom to insist that experience is also connotative. More, as Dewey stresses, all inquiry, i.e., all reflective experience of nature, has an experimental potential. Inquiry is not limited to naming and placing, for it can forge, shift, and reconstruct the way in which our experiences come at first-hand. At an initial glance this seems to make considerable sense as a description of our basic transaction with the world. Alas, the description turns out to be far more ideal than characteristic.

Two major obstacles loom in the way of our capacity to enter into an experimental, resolving, healing and future oriented transaction with the world as it comes to us. First, as Dewey stresses over and over, we tend not to trust our own experience to be significant, leading, warning and revealing. In fact, we have been taught to either deride the importance of our own experience, and therefore, enter an incommunicative shell, or from insecurity and a hesitant sense of the true worth of our experiences, we rattle on in a monologue, as self-aggrandizing as it is empty. Second, in our confrontation with our own experience, we are all too often severely hobbled by the trappings of our own education, informal and formal, especially the latter. Although the world as experienced speaks in dulcet, loud, peremptory, cajoling and symbolic ways, it is rare that we hear anything. Although the world as experienced struts its colors, carried by blossom, bird and sky, it is rare that we see anything. Although the world as experienced preens its scents, floated by sea, cave, animal and cuisine, it is rare that we smell anything. Although the world flaunts its texture by bristle, horsehair, cactus and spider web, it is rare that we feel anything by touch. It is not as Yeats warned, that the center does not hold, rather it is that the senses do not grasp, do not reach, for they have become captives of the denotative mind rather than extensions of our bodies.

Given the riot of possibilities in the ancient world, who can fault Aristotle for putting "things" and "species" in order, according to his own light. Before Aristotle, in the writings of Plato, and since Aristotle to this day, the poets have been struggling to free our experience from the clutches of definition and the roster of self-defining names, which have rendered our experience more of the inert than of the symbolical. No one says this more perceptively than Dewey's mentor, Ralph Waldo Emerson.

> We learn nothing rightly until we learn the symbolical character of life. Day creeps after day, each full of facts, dull, strange, despised things, that we cannot enough despise,—call heavy, prosaic and desert. The time we seek to kill: the attention it is elegant to divert from things around us. And presently the aroused intellect finds gold and gems in one of these scorned facts,—then finds that the day of facts is a rock of diamonds; that a fact is an Epiphany of God.

Returning to the Stoics, they push farther. Aurelius tells us that it makes no difference how long we live, for life is a warfare and a stranger's sojourn. Yet, our bond is holy, sacred, for nature needs us, eats us. We are the stuff, dead though we be, of the future. *Quo vadis*. Into the future our dead bodies are eaten by a subsequent generation, who in turn will be eaten. The Greek and Roman cycle makes this palatable. For Whitman, post-Copernican, the rhythm continues, unabashed. The *Leaves of Grass* are the uncut hair of the dead. The conclusion of the poem, camarado, is that after our work together, I go off, by myself. Come look for me, and you shall find me, under your boot soles.

There is a thickness here, one in stark contrast with the banality, the emptiness and the conceptual aridity so endemic to the commentator on matters epistemological. Think only of the difference between Marcus Aurelius and G.E. Moore.

Just what is my position on this issue of how we come to be, who we be, who we are, who "they" think we are, and just what we are "up to," that is, how it goes with us, persons in a hostile, unrelenting, unforgiving and transient eco-system. Or, in a paraphrase of Kant; what, what in the name of all that is sacred, can we know?

I see no reason to fool around, to make out as though we belong, as though "it" makes sense, as though we have a place, a natural place. I see no place for us, cosmically, even planetarially, in fact, I have painful doubts about whether we have even a local place. Aristotle, genius though he be, sold us a bill of goods, one, ironically, taken up by modern Christendom. Plato was wrong but he did have a sense of the disaster awaiting us, so he invented a secret, a way out. Aristotle, stodgy and literal, denied the possibility of a secret, correctly but unfortunately, and said that we and the world cohere. The congeniality of self and world has misfired. Information becomes ontology. Phusis, the harbinger of salvation, the source of *logoi spermatikoi, rationes seminales*, becomes in a macabre dealing of the deck, a hand of twos and threes, subject to a flush, *out*.

I stand here, talking to you. For me, the history of the human self, actual and rhetorical, is as real, as palpable as your body. The opponents weigh upon me. My fragility cries out in desperation. To whom, to what, do I turn? I have some collegial friends from the recent past. Let us try them as a wedge into a possibility of liberation from the overwhelming evidence that tells me I am trapped, trapped, for ever more.

William James says that the will is cognitive. This is an extraordinary claim. If I will to believe that I make a difference, then, it is possible that I make a difference. James assures us that there is no guarantee in such a claim, yet he assures us that more often than not, the difference is due to our will to believe. Furthermore, James contends that our very selves are dependent on our choice, on our *willingness* to do, suffer and create.

Most commentators focus on the promethean character of the Jamesian self. I, however, note its fragility. No one will be self–no self. I repeat. *No* self. We, I, You, become evanescent, a spook, a shadow, a schmoo left behind the raging torrents of obviousness. Show up. Be someone, someplace. Cannot I just be. No!

No? Sartre and Camus say we can just be. Or do they? On a closer glance, perhaps note that the existential self has been a source of temporary liberation for those of us who have come to consciousness in the era of *Das Man*. Choose! Deny! Assert! Strike off those shackles of yesterday. Be free. Be Mersault of Camus' *L'Etranger*. Do not let them "exacerbate" you. Peel your own skin. Die while saying hello.

What then am I to do? The closet of the ancients, brilliant, but of another world, another time. The exquisite composition of Descartes with its price, my body. The promethean self of Kant and William James, stoved in by the merciless message of Marx and Freud. I seek among the rubble for the shards, lonely shards, of a new start. John Dewey tells me that the Live Creature is not a redundancy. He tells me that I should live like the animals, on the *qui vive*. I think of my Irish setter, Shannon by name, whose nostrils gathered the world and called upon him to spring from his quiescent haunches in pursuit of a smell, a sound, a presence. Dewey tells me that how I feel, how I touch, how I reach, how I fail, all counts in this persistent attempt of ours to seek from an obdurate nature, something worthwhile, sometime okay, someday, somewhen now, somehow this, somewhere here, as in Whitman; 'hey there,' you, camarado, I love you, why do not you answer and so, I say again 'hey there,' 'hey there.'

I return; what then am I to do? Wounded and confused, I seek to make a new beginning. So, how goes it with me? What am I to do, so that I be!? My first task is to deal with my inheritance, genetic, familial, ethnic, cultural, regional, political, religious and stripe of maturation.

Inheritance, both genetic and cultural, is a many-splendored, pock-marked bequest. Although Marx was theoretically right, he was shy in his perception of the influences which detail our coming to consciousness. Contemporary anthropology has made it painfully clear that we inherit virtually all that we come to be. Gait and gesture, sensibility, assumptions, anticipations, self-diagnosis are all consistently and persistently trapped in the mimetic which accompanies burgeoning awareness.

No later than adolescence do we move off the mark already heavy laden with the grave but present bequest of millennia past, for example, the Chinese. More pressing is the weight of our western strand, Abraham and Moses, succeeded by the umbrella of things Greek and Roman. Soon thereafter, Jesus points a finger at us, while at a distance, now closing fast, Islam camels its way into the Pyrennees. Modern science, induction, the experimental method burst upon us as Kings, Queens and Squires become an antiquity. The land of the *mundus novus* beckons and machines grind their way into the fabric of our lives. The riot, plaid and controverted, of ethnicity claims each of us for one or the other. Urban, rural, west, south, east, north, river, ocean, mountain, plain, desert, abode, adrift, shack, and places galore, each tethering us to the world as we struggle to be something while of, about and in something.

Can it be that the I which intends to be me in this turbulent sea of 'other,' can really be me as different from you? Ah! the fingerprint is unique and the toeprint also. The face claims univocity. Isolated events happened to me, not to you and to you, but to me. We hold them as signs of 'me' stuff rather than 'you' stuff, a sort of wail in the blizzard, an attempt to match the genius of the eskimos who have more than ninety words for the snow which we call by one word. Pressed, however, this doesn't do. Novelty is not external. Novelty demands an affective, personal, hey there, look at me. Did you hear what happened to me? Now, I think! This is my opinion! If I were you! I can't get started! I ain't nobody. I think I need help! Billie Holiday sings, "Why was I born?" The plaintive and desperate answers, "to go on loving you." But "you" do not love me!

In all of these remonstrances, beggings, assertions the I, which is me, breaks through the cloud of anonymity, gathered only by the context and fulcrum of situation. No situation, no one, no me, no I. Not I, no you.

Duplication, triplication, quadruplication, the power to N, surrounds us. Faced with this, how do I say my name? Only by socializing my experiences, for they, if I claim them, are mine and only mine. Only by carrying my personal history, effluvia for others, but for me, stuff of my stuff. Only by reconnoitering my past and marrying it to my present. Only by incessantly making relations. Only by busting every stereotype. Only by saying *that* is not how I see it, feel it, hear it or be it. Only by sucking originality out of the most mundane, prosaic events. Only by rendering every dot, every fact, every happenstance, every repetition, a symbolic event.

So much for inheritance. A noose, a burden, yet a spring board from which a lurch. As I see it, touch it, smell it, hear it, however furtively, the task is to seize, build, construct, create, the time, the day, the space, the place, the self. Yes, I believe that we have no distinctive self as inherited. Yes, I believe that the soul, my soul, your soul, is a mock-up. Soul, me, I, person, is a self-generated construct. Our lot, genetically, historically, traditionally, inheritedly, is to be but a ghoul, alive, but experientially dead, a *manque*, merely a shadow, a wizard of Oz, pumping sounds and *accountrement*, without any presence, any self, any answer to who goes there.

Hey there, just what do you mean by epistemology, by knowing, by knowing something, the world? Allow me to tell you about the cigar man.

Long dead, long ago on a dreary Sunday morning, he taught me about the difference between the literal and the symbolical. In 1948, I was a volunteer at the *Catholic Worker*, in New York City headed by the indefatigable and famous Dorothy Day. One of my tasks was to visit hospitals and attend to the indigent sick. On early Sunday mornings, I would go to Bellevue Hospital and do what I could to assuage the suffering in the cancer ward, pronounced as the death ward, because this was before the contemporary therapies were available. Among my assignments was the pathetic effort to fit the patients with appropriate reading glasses, collected randomly at the *Worker*. I also made phone calls, ran errands and assisted the crippled patients to chapel services. One rainy Sunday morning, I came upon a quadruple amputee, sitting totem-like on the bed, as if he were a runaway torso. He was Puerto Rican and his black eyes had the defiance known only to a beaten person, for he was terminal. I passed by his bed, at a quickened pace, afraid to look or stare. He called to me in Puerto Rican. Not knowing the language, I asked a nurse for help. Our patient wanted two cigars. It was a rainy Sunday morning and not propitious for finding a store open. I groused but said I would try to get them for him.

As I was about to leave, he barked again to the nurse. He wanted a special cigar, a Puerto Rican cigar. At that point, I became irritated for the task was becoming onerous. I told the nurse to tell him that I would try for his cigars, but failing, I would bring him some American cigars. Upon being told that, he became furious and all of his anger, as befits a quadruple amputee, rushed to his face. He blurted: no Puerto Rican cigars, no cigars at all. As embarrassing as it now seems to me, I became irritated at his obduracy. After all, I was doing a "good thing," being helpful and solicitous. But enough is enough. Yet, as an Irish Catholic, I had been steeped in guilt and in the moral obligation of service, so I set out to get two Puerto Rican cigars, no easy task. After much hassle, I did obtain the cigars and brought them back. I placed one in his mouth and lit it. His face shone like an evening star.

It was not until fifteen years later that I realized what had happened, what I had missed and what I had to learn. I was teaching the philosophy of Albert Camus and the theme was personal authenticity. In a flashback, my Puerto Rican patient taught me of authenticity, of the aesthetic moment and of the quality of life in a pretechnological medical setting. His message came to me as a revelation, an epiphany. Surely, only those special cigars would do, for only the smell and taste would bring back the rich memories of Puerto Rico, the cocoa-brown skin of the women, the mist hanging over the coast between the Condado of San Juan and the city of Mayaguez. Only those cigars were authentic, existential, freighted with experienced relations. Only they sang his song, his personal song.

In my carrying on, our deepest ontological state is that of transiency, a passing through and out, out of being. We live the *apeiron* of Anaximander without the comfort of repetition, the historical cycle. If there be a nectar for us, a source of nutrition, it is to be found in the journey, striated by hope, despair, and the persistent effort to live on the cutting edge where experience flowers, teaches, chastises.

The entire effort of personing is both literally and figuratively in our hands. We knead, we shape, we caress, we fist and we warn. Above all, we reach, like Shelley's quest of the moth for the star. And like the moth, we hover about a glimmer bathing for a time in the light, knowing full well that the last word is that of the shadows which precede the gathering darkness. I am with Camus. I want to live with what I know, what I experience and not with a nostrum, fake and deceiving. Both world and person are mock-ups. The quality of our life is the attendant care with which we build out personal space, our place, in the throes of time, passing. For every hello, it is inevitable that we also say, good bye! Meanwhile, where, what did you do with my things?

NOTES ON CONTRIBUTORS

Albert Borgmann (Ph.D., University of Munich, 1962) teaches philosophy at the University of Montana and is the author of *The Philosophy of Language* (1974) and *Technology and the Character of Contemporary Life* (1984). He is now at work on a book about the end of the modern era.

Ronald C. Bruzina (Ph.D., University of Notre Dame [Indiana], 1966; Doctorat de 3 cycle, Universite de Paris-Nanterre, 1970), is Professor of Philosophy at the University of Kentucky. He is the author of *Logos and Eidos: The Concept in Phenomenology* (1970) and of numerous essays in phenomenology and the philosophy of technology. He co-edited (with Bruce Wilshire) *Phenomenology: Dialogues and Bridges* (1982), and *Crosscurrents in Phenomenology* (1978), and translated Kostas Axelos, *Alienation, Praxis, and Techne in the Thought of Karl Marx* (1976). Currently recipient of a research grant from the National Endowment for the Humanities, he is preparing a book on the contribution of Eugen Fink to the development of phenomenology in Husserl's late philosophy, as well as a translation of the recently published *VI. Cartesianische Meditation*, also by Eugen Fink.

Timothy Casey (Ph.D., Duquesne University, 1986) is Assistant Professor of Philosophy and Director of the Center for Science, Technology and Society at the University of Scranton. He has published essays in philosophy of technology and phenomenology and is especially interested in the relation of technology to architecture and history.

Edward W. Constant II (Ph.D., Northwestern University, 1977) is Associate Professor of History at Carnegie-Mellon University. His publications include *The Origins of the Turbojet Revolution* (1980). He has an abiding interest in the relationship between science and technology, and in the epistemological foundations of scientific and technological knowledge and practice.

Lester Embree (Ph.D., New School, 1972) is Professor of Philosophy at Duquesne University and President of the Center for Advanced Research in Phenomenology, Inc. He has edited and translated a number of books and authored various essays in the history and philosophy of technology as well as the human sciences and in constitutive phenomenology. He is currently preparing *The Rise of American Theoretical Archeology.*

Frederick Ferré (Ph.D., St. Andrews University, Scotland, 1959) is Research Professor of Philosophy at the University of Georgia. He is author of *Language, Logic and God* (1962), *Basic Modern Philosophy of Religion* (1967), and *Shaping the Future* (1976). His most recent book is *Philosophy of Technology* (1988), and he combines interests in philosophy of religion and ethics with philosophy of technology and environment. He is editor of *Research in Philosophy and Technology.*

Thomas R. Flynn (Ph.D., Columbia, 1970) is Samuel Candler Dobbs Professor and Chair of the Department of Philosophy at Emory University. He is the author of *Sartre and Marxist Existentialism: The Test Case of Collective Responsibility* (1984) and is currently completing a book on Sartre, Foucault and reason in history.

Larry Hickman (Ph.D., The University of Texas at Austin, 1971) is Associate Professor of Philosophy at Texas A&M University. His publications include *Modern Theories of Higher Level Predicates* (1980), *Technology and Human Affairs* (ed. with A. al-Hibri, 1981), *Philosophy, Technology and Human Affairs* (ed., 1985), *Technology as a Human Affair* (ed., forthcoming 1989), and *The Culture of Instrumentalism: John Dewey's Pragmatic Critique of Technology* (forthcoming 1990).

Ivan Illich is researcher in the Science, Technology and Society Program, Penn State University, and in the Center for Intercultural Documentation, Cuernavaca, Mexico. He is the noted author of *Tools for Conviviality, Medical Nemesis*, and, most recently with Barry Sanders, *A B C: The Alphabetization of the Popular Mind.*

John J. McDermott (Ph.D., Fordham University, 1959) is Distinguished Professor of Philosophy and Humanities and Head of Humanities in

Medicine at Texas A&M University. He has authored numerous books and essays on pragmatism and philosophy of technology.

Algis Mickunas (Ph.D., Emory University, 1969) is Professor of Philosophy at Ohio University. He has coauthored, coedited, and translated various works, has published articles in diverse disciplines, from philosophy through sociology, pedagogy, communications, comparative civilizations, semiotics, and film.

Carl Mitcham, (Ph.D., Fordham, 1988) is Director of the Philosophy and Technology Studies Center at Polytechnic University (New York). His books include *Philosophy and Technology* (1972, 1983), *Bibliography of the Philosophy of Technology* (1973, 1985), *Theology and Technology* (1984), *Philosophy and Technology II*, (1987), and *¿Qué es la filosofia de la tecnología?* (1989).

Gayle L. Ormiston, (Ph.D., Purdue University, 1981) is Associate Professor and Chair of Philosophy at the University of Colorado at Colorado Springs. His publications include *From Artifact to Habitat: Studies in the Critical Engagement of Technology* (1989), *Narrative Experiments: Interpreting the Labyrinths of Science and Technology* (1989, with Raphael Sassower), *The Hermeneutic Traditions: From Ast to Ricoeur* (1989, with Alan D. Schrift), *Transforming the Hermeneutic Context: From Nietzsche to Nancy* (1989, with Alan D. Schrift), and numerous articles on recent French and German thought.

Joseph C. Pitt (Ph.D., University of Western Ontario, 1972) is Professor of Philosophy, Adjunct Professor of Science Studies and Director of the Humanities, Science and Technology Program at Virginia Polytechnic Institute and State University. His main interests are in the history and philosophy of science and technology. The author of *Pictures, Images, and Conceptual Change* (D. Reidel, 1981), numerous articles, and the editor or co-editor of nine volumes in history and philosophy of science and technology, most recently *Theories of Explanation* (Oxford, 1988), he is currently working on two book length manuscripts, *How to Think About Technology* and *The Book of Nature: Galileo and Commonsense.*

Robert N. Proctor (Ph.D., Harvard University, 1984) is Coordinator of the Program in Science, Technology and Power at Lang College, The New School for Social Research. His publications include *Racial Hygiene: Medicine Under the Nazis* (1988) and a forthcoming book on the origins of the ideal of value-free science. He is also preparing a more general essay on the political history and philosophy of science.

Wolfgang Schirmacher (Dr. Phil. Universität Hamburg/West Germany, 1980) is Research Associate Professor of Philosophy in the Philosophy and Technology Studies Center at Polytechnic University, as well as Faculty Member in the Media Studies Graduate Program and in Social Sciences at the New School for Social Research, President of the International Schopenhauer Association, Editor of Schopenhauer-Studien, and Program Director of the International Heidegger Congress 1989 in Paris. His publications include *Ereignis Technik* (1980), *Technik und Gelassenheit* (1983), *Schopenhauer* (1985), *Zeitkritik nach Heidegger* (1989). He has published essays on 19th and 20th century continental philosophy with emphasis on phenomenology, existentialism, bioethics, philosophy of technology, and media philosophy.

Langdon Winner (Ph.D., University of California, Berkeley, 1973) is Associate Professor of Political Science at Rensselaer Polytechnic Institute. His publications include *Autonomous Technology* and *The Whale and the Reactor*. He is President of the Society for Philosophy and Technology and is currently working on a book on political theory and technological design.

INDEX

aesthetic differentiation, 246, 255
aesthetic values, 165, 169-70, 172, 175-77, 179
age of empire, 8
Alberti, Leone Battista, 219
Albertus Magnus, 89
altar candle, 239
Anaximander, 301
anti-Semitism, 20, 37
applied science, viii, 151, 261
Aquinas, Thomas, 91, 229
architects, 202-3, 205, 215, 218-19, 244, 246-51, 253-54
architectura architecturans, 204-6, 218
architecture, 96, 118,
  as sculpture, 249
  beauty, 244-46, 248-50, 254
  city and, 210, 212
  cultural context and, 203
  definition of, 205, 207, 219, 244-45
  failure of, 244
  forces in, 253, 257
  form, 214, 216-17, 243-45, 250-57
  function, 213-14, 216-17, 243-46, 248-55
  history of, 201, 203
  metaphoric nature of, 254
  modern, 244-49
  philosophy and, 243-45
  postmodern, 249
  religion and, 238-39, 257
  symbolic meaning of, 217, 226, 256
  technology and, 245-46, 252
  Vitruvian elements in, 248
Aristotle, vii, 8, 171, 183, 147, 229, 293-97
artificial intelligence, 115-16
Augustine, 229, 294
Aurelius, 296-97
authenticity, 301
automation, 5, 109-10
Barnes, Hazel E., 287
Beda Venerabilis, 90
being-in-the-world, 290
belief, 153, 297
Benjamin, Walter, 124-25, 130
Bernard of Clarivaux, 75, 82, 85, 92
Bible, 82-83
bicycle, invention of, 183-84
Bijker, Wiebe, 183, 197-98
Billington, David P., 199
biologism, 23, 40, 46
biomedicine, 19-24, 40-45, 300-1
Bloom, Allan, 119-20, 126-27, 130-31, 142
Boethius, 229
Bonaventure, 79
Borgmann, Albert, 119-20, 126-27, 130, 282-83, 287
Boulée, Etienne-Louis, 245, 254
Braudel, Fernand, 148, 163
Brown, Norman O., 294
Bruzina, Ronald, 179
Burian, Richard, 179
Butterfield, Herbert, 183, 198
buying a car, 157
Cairns, Dorion, 163
Callon, Michael, 184, 197
Campbell, Donald, 185, 191, 197-98

Camus, Albert, 293, 298, 301
candle-making, 239
capitalism, 3, 57
Carr, E.H., 183
Carrel, Alexis, 35
Carver, Norman F., Jr., 256-57
Cassiodorus, 80, 90
Chadwick, John, 113
Christianity, 290, 297
Clanchy, M.T., 84
clothes, 222-24, 237-38
cognitive values, 165, 169
common sense, 294
Common Sense Principle of Rationality (CPR), 177-79
Cummings, Robert D., 287
communal practices, 98-100, 102, 104
compilations, 72, 76, 86
computers, 4, 12-13, 15, 70, 72, 115-16
Condit, Carl W., 199
Constant, Edward W. II, 199
convenience,
conventionalism, 183
corporate vs. local culture, 10-11
craft technology, 149
Dante, 91
decision procedure, 174
deep ecology, 9, 15
"deformation," 252, 257
de Havilland, Geoffrey, Jr., 191
Deleuze, Gilles, 136, 142, 267-69
democracy, 18, 49, 53, 58-59, 62-63, 111
Derrida, Jacques, 139, 141, 143, 257, 259, 268
Descartes, Rene, 55, 67, 126, 293, 298
destiny, 279-80
Dewey, John, 128, 131, 265-66, 268-69, 285, 294-95, 298
dialectical reasoning, 284
Dickens, Charles, 128
difference, 133, 139, 251, 257, 259, 292, 299
Durand, Jacques-Nicholas-Louis, 245-46
ease of use, 253
Einstein, Albert, 84
Eisenstein, Sergei, 128, 131
electronic space, 1, 3-4, 6, 12-14
electronic trading, 6
Eliade, Mircea, 206, 219
Elzen, Boelie, 184, 198
embodiment, 294
Embree, Lester, 84, 163
Emerson, 290, 296
empirical knowledge, 170, 172
empirical program of research (EPOR), 182-84, 197
ends and means, 152
enlightenment,
  political, 49-53, 57, 59, 65
  scientific, 49, 53-57, 60, 64-65
Epictetus, 8
epistemic values, 169-72, 175-77, 179
epistemology, xi, 158, 300
equality, 52
equipment, ix, 146, 151-56, 162
ethical values, 169, 172
Eucharist, 233-35
eugenics, 20-22, 24, 29, 57
euthanasia, 29, 35-39
evaluation, 152, 161-62
"evidencing," 158
exigency, 278-80, 287

existentialist hope, 285
experience, 294-96, 301
faith and reason, 237, 239-41
fascism, 18, 45
feminism, 20
Ferguson, Henry, 10-11, 16
film, 118-19, 121-22, 124-25, 127-28, 134, 136
Fink, Eugen, 50, 66
folk architecture, 256
Forster, E.M., 127
Foucault, Michel, 272, 286
Freud, 298
functional dimension, 253
Gadamer, Hans-Georg, 246, 254-55, 257
Galileo, 54, 171, 180
Ganz, David, 89
genetic inheritance, 298-300
genocide, 45
Giedion, Sigfried, 254
Gomez, Alberto-Perez, 256
Goodman, Nelson, 112, 180
Guttenberg, 74
Hall, A.R., 260, 268
Haugeland, John, 112
Hegel, 61, 285
Heidegger, Martin, viii, 140, 143, 269, 271-72, 286
Heraclitus, 294
Hesse, Mary, 183, 190, 198
"heterogeneous engineering," 184, 187
historians, 69-70, 183
historical materialism, 280
historicism, 192
history, 281-82, 285
Hitler, 23-24, 32-33
Holmyard, E.J., 260, 268
Homer, 3
Hook, Sydney, 126, 131
Horace, 95-96, 112
Hughes, Thomas P., 184, 197
Huidobro, Vicente, 69, 84
Hugh of St. Victor, 70-73, 78, 83, 86
human autonomy, 51-54, 56-57, 61
Husserl, Edmund, viii-xii, 152, 271, 273
Ihde, Don, xii
Illich, Ivan, 84-85
information age, 3, 111
institutions, 251, 257
interest, 279
International Style, 244, 248, 254
Islam, 299
James, William, 293-95, 297-98
Jews, 20, 32-33, 35, 38, 42, 45, 47
Johnson, Philip, 243, 247
Jonas, Hans, 56-67
Joyce, James, 121-22
Kahn, Louis, 250-53
Kant, 297-98
Kidder, Tracy, 247, 256
Kostof, Spiro, 105, 219, 247, 256-57
Kriele, Martin, 52, 66, 68
Kuhn, Thomas, 17-18, 180-82
Landgrebe, L., 61, 67
Law, John, 184, 197-98
Le Corbusier, 214, 216, 220, 247-48
Leonardo da Vinci, 190
Lewis, S., 91
liberalism, 58-63
life project, 272, 274-75, 290
lifeworld, viii-ix, xii, 5, 133-34, 136, 139-41, 246-47, 250, 253-54, 272
literal and symbolical, 300-1

literacy, 94, 97-98, 111, 118-19, 129-30
Locke, John, 7, 15
logocentrism, 139
Logos, 294
Long, Marion, 111, 116
loneliness, 290-92
machines, 150, 271, 278-79
MacLean, Norman, 105, 114
Marcuse, Herbert, 294
Margolis, Joseph, 118, 130
Mark, Robert, 199
Marx, Karl, 53, 61, 122, 280, 282, 298
Marxism, vii
Marxism-Leninism, 57
McLuhan, Marshall, 133, 142
media,
  as autobiography, 133-34
  as lifeworld, 133
memory, 231-34, 293
Mengele, Josef, 27
Merleau-Ponty, Maurice, xii, 134, 142-43
Mirandola, Pico della, 51
Mitcham, Carl, xii, 142, 237, 239-40, 286
modern ontology, 50-52, 61, 64
modified technologies, 240
Moore, G.E., 297
movement-image, 136-37
Mumford, Lewis, xii, 147, 250, 256, 279
music, 2, 96, 106-7, 114-15, 118, 121, 130, 134-41, 227-31, 250, 299
mythology, 206-7
National Socialism, 17-48
  abortion under, 32
  doctors and, 23-28, 39-43
  philosophy and, 45
  science and, 19, 27, 42-46
  women and, 30-33
natura naturans, 204
nature, viii, xi, 53-56, 146, 290, 293-94
neo-Stoicism, 9, 293
Nietzsche, Friedrich, 53, 205, 219, 230
nihilism, 295
Noble, David, 108-10, 115-16
non-specialized technology, 148
nostalgia, 292-93
Nuremberg Laws, 33-35
objectivity, 170, 177
obsolescence, 293
Ong, Walter, 84, 112-14
Parkes, M., 86, 88
Pascal, 183
perception, 139, 153
Perrault, Claude, 245-46, 248
Peter Lombard, 73-74, 85-87
phenomenology, viii-ix, 134-36, 151, 153, 221, 237, 241, 284, 293
"philosophic fallacy," 128
philosophy of science, x
philosophy of technology, vii, x,
phonetic writing, invention of, 76
photography, 124-25
place, 201, 205-10, 212-13, 218-19, 244, 253, 290-92, 297, 299-301
Plato, vii, 63, 119-20, 147, 229, 290, 293-94, 296-97
play, 22
Plotinus, 229
poets, 296
polis, 8
politics, 7, 13-14, 18, 43-44, 49
Pollard, Graham, 92
Popper, Karl, 17, 18, 45
Portoghesi, Paolo, 251-52, 257

Postman, Neil, 133, 142
Powers, Richard, 124-25, 131
practical attitude, 147, 271, 273
practical life, viii-ix, 243
practico-inert, 277-78, 280-82, 284-85
pragmatism, vii, 285
praxis, 277-78, 280-84
presencing, 137-38
process, 278
production and consumption, 155, 162
Prometheus, 239
property, 7
racial hygiene, 19-20, 23-24, 27, 30, 33-35, 40, 45
reading,71-72, 98-102, 113, 118
    memory and, 78, 81, 103, 119
    monastic, 71, 73, 75, 80
    scholastic, 71, 75
Reagan, Ronald, 60
reason, 156, 160
recombinant DNA research, 166-68, 173-76, 178, 180
reflection, 151
religious consciousness, 237
Richter, Horst, 50, 66
Rouche, M., 84
Rouse, R.H., and Rouse, M.A., 85, 88
Rudner, Richard, 169, 180
Rudofsky, Bernard, 220
Rybczynski, Witold, 256-57
Ryle, Gilbert, 175
Sallust, 89
Sartre, Jean-Paul, 271-87, 298
Sartrean freedom and responsibility, 272, 276, 280, 285-86, 298
scarcity, 277-78, 280-83, 290-91
Schabert, Tilo, 50, 54, 66-68
Schopenhauer, 230
science, ix, xii, 159, 184
    ideology and, 17-19, 40, 42-44, 46
    sociology of, 182
    value judgments and, 167
scientific change, 18, 62
scientific technology, 145, 150-51, 185
Scripture, 231-34, 294
Shelley, 301
Signer, Ethan, 166-67, 173, 178
Sinshimer, Robert, 167
Singer, Charles, 260, 268
situation, 274, 285
Sloterdijk, P., 65, 68
Smalley, B., 85, 88
Smith, E. Baldwin, 220
Social Darwinism, 19, 23
sociobiology, 43, 48
Socrates, 221
space, 138, 206-12, 230, 246, 291-92, 300-1
Spaemann, R., 62, 67-68
spatial location, 1, 3-5, 7-8
specialized technology, 149
Spinoza, 204-5, 219
St. Paul, 230
sterilization, 29-30, 33, 38, 46
Stoics, 293, 296
"strong idea," 251, 253
Suetonius, 89
suicide, 291
suitability to use, 253
symbolic logic, 101-2
Synode of Arras, 90
Talbot, C.H., 92

Taylor, Frederick Winslow, 13
technocracy, 44, 57, 59-65
technological age, vii, 134, 241
technological artifacts, 289
technological change, 181, 190, 192
technological culture, vii, x-xi
technological decision-making, 176-79
technological determinism, 126, 130, 280
technological lifeworld, 136
technology, ix-xi, 57, 65, 93, 134, 141, 191, 272, 275, 279
- alienation and, 271, 273, 276-77, 280, 283, 290
- alphabet as, 69-71, 76-78
- architecture and, 245-46, 252
- as neutral tool, vii-viii
- business ethics and, 11-12
- church and, 221-36
- critique of, 161
- definition of, 146, 165, 238-39, 260-67, 277, 286
- disburdenment and, 110
- ethics and, 166-67, 173
- habitat and, 262-64, 266-68
- history of, 260-62, 266
- liberation and, x-xi, 49, 52, 57-63
- National Socialism and, 19
- progress and, 56, 62-64, 282
- questioning and, 259-63
- religion and, 238-41
- risk and, 168
- science and, viii-ix, xii, 54-56
- semiotics of, 283-84
- social construction of (SCOT), 182-84, 187-90, 192, 197
- text as, 74-75, 77-79, 82-83, 93-131
- translation and, 264-65,
- values and, 168-69

teleconferencing, 2, 4, 14-15
television,121
Tertullian, 9
texts, 117
- as pure appearance, 106-7
- as tools,123
- canonical, 115
- literary, 94-96, 104, 118, 121-22, 126, 128
- realization of, 95, 97, 101, 104-5, 108, 110-11, 118
- technological, 93-94, 97, 105-12, 117-19, 123
- traditional, 93-95, 97-98, 102-3, 106, 111, 117-19, 123

text-type determinism, 117, 120-22, 126 129
things, 289-91, 293, 301
time, 137-38, 210-11, 230, 275, 292, 301
Todd, William, 200
transnationals, 9-10, 12
transportation, 225
turbojet,
- invention of, 181-200
- inventors of, 185-87
- piston aircraft engines and, 186-87
- reasons for invention of, 188, 191-92

ultracentrifuge, 184
use character, 152
Venturi, Robert, 252
Vico, 230
videos, 107, 122, 133-41
Vincent of Beauvais, 86

violence, 281-82, 287
volition, 152, 162
Vonnegut, Kurt, 109, 113
Wagner, Richard, (*Gesamtkunstwerk*), 136
water, 291
White, Lynn, Jr., xii, 145, 150, 163
Whitman, Walt, 289, 296-98
will, 297
Willy Loman, 293
Wittgenstein, Ludwig, 260, 268
world, 201, 207-16, 253, 255, 273, 276, 291, 293, 295-98, 301
world appropriation, 100, 271-72, 275
worship, 227-31
Wright, Frank Lloyd, 247
Yeager, Chuck, 191
Yeats, W.B., 296